Mars via the Moon
The Next Giant Leap

Erik Seedhouse

Mars via the Moon

The Next Giant Leap

Published in association with
Praxis Publishing
Chichester, UK

Erik Seedhouse
Assistant Professor, Commercial Space Operations
Embry-Riddle Aeronautical University
Daytona Beach, Florida

SPRINGER-PRAXIS BOOKS IN SPACE EXPLORATION

Springer Praxis Books
ISBN 978-3-319-21887-8   ISBN 978-3-319-21888-5 (eBook)
DOI 10.1007/978-3-319-21888-5

Library of Congress Control Number: 2015957057

Springer Cham Heidelberg New York Dordrecht London
© Springer International Publishing Switzerland 2016
This work is subject to copyright. All rights are reserved by the Publisher, whether the whole or part of the material is concerned, specifically the rights of translation, reprinting, reuse of illustrations, recitation, broadcasting, reproduction on microfilms or in any other physical way, and transmission or information storage and retrieval, electronic adaptation, computer software, or by similar or dissimilar methodology now known or hereafter developed.
The use of general descriptive names, registered names, trademarks, service marks, etc. in this publication does not imply, even in the absence of a specific statement, that such names are exempt from the relevant protective laws and regulations and therefore free for general use.
The publisher, the authors and the editors are safe to assume that the advice and information in this book are believed to be true and accurate at the date of publication. Neither the publisher nor the authors or the editors give a warranty, express or implied, with respect to the material contained herein or for any errors or omissions that may have been made.

Front cover image of astronaut used under license from Shutterstock.com
Cover design: Jim Wilkie

Springer International Publishing AG Switzerland is part of Springer Science+Business Media (www.springer.com)

Contents

Acknowledgments	ix
Dedication	xi
About the Author	xiii
Acronyms	xv
Kick-Starting the Next Giant Leap	xvii

1	**Martian Dragons**	1
	Dragon #1: radiation	2
	Dragon #2: cataracts	9
	Dragon #3: blindness	12
	Dragon #4: bone loss and muscle atrophy	13
	Dragon #5: early-onset alzheimer's and brain damage	17
	Dragon #6: entry, descent, and landing	18
	References	25
2	**Government Anchors**	27
	Lunar nations: Russia	28
	Lunar nations: China	35
	Phase 1: orbiting the Moon	38
	Phase 2: landing on the surface	38
	Phase 3: sample return	39
	Phase 4: boots on the Moon	39
	Lunar nations: European Space Agency (ESA)	40
	SinterHab	43
	Lunar nations: US	45
	References	56

Contents

3 Commercial Anchors ... 57
   Moon open for business ... 58
   Lunar markets ... 58
   Resource extraction ... 59
   Tourism ... 60
   Technology, test, and demonstration ... 61
   Defense and security ... 63
   Science and exploration ... 65
   Education ... 65
   Support and supplies ... 66
   Media ... 66
   Shackleton Energy Company (SEC) ... 66
   Shimizu Corporation ... 70
   Bigelow Aerospace ... 71
   OpenLuna Foundation Inc. ... 73
   Golden Spike ... 74

4 Technology ... 79
   Transport ... 80
   Hazard avoidance ... 85
   Life-support system ... 86
   Habitat design ... 92
   Robotics and rovers ... 93
   Structures and construction ... 98
   In-situ resource utilization ... 100
   Communications ... 106
   References ... 107

5 The Human Element ... 109
   Selection ... 109
      Pre-emptive surgery ... 111
   Sterilization and related issues ... 113
   Genetic testing ... 114
   Getting along ... 114
      Uncharted territory ... 116
   Martian analog mission ... 120
   Death sims ... 123

6 Regulation ... 125
   Moon Treaty ... 126
   Property rights ... 129
   Salvage ... 132
   Commercial operations and Rules of the Road ... 134
   Pollution ... 135
   References ... 136

7	**Headwinds and Tailwinds**...	137
	Headwinds...	138
	Grounded..	138
	Financial crisis	140
	International Lunar Decade..	140
	Space race	143
	Asteroid...	144
8	**Making the Moon Pay: The Economical and Logistical Viability of Boot Prints on the Moon**	147
	Mining...	152
	Platinum..	155
	Propellant..	156

Appendix I 161

Appendix II. International Lunar Decade Goals 163

Epilog 165

Index... 169

Acknowledgments

In writing this book, the author has been fortunate to have had five reviewers who made such positive comments concerning the content of this publication. He is also grateful to Maury Solomon at Springer and to Clive Horwood and his team at Praxis for guiding this book through the publication process. The author also gratefully acknowledges all those who gave permission to use many of the images in this book, especially Shackleton Energy Company.

The author also expresses his deep appreciation to Christine Cressy, whose attention to detail and patience greatly facilitated the publication of this book, and to Jim Wilkie for creating yet another striking cover. The author would also like to thank Production Manager, S. Reka, for her help in bringing this book to publication.

To those who eventually take the next giant leap

About the Author

Erik Seedhouse is a Norwegian-Canadian suborbital astronaut. After completing his first degree, the author joined the second Battalion, the Parachute Regiment. During his time in the "Para's" Erik spent six months in Belize, training in the art of jungle warfare. He also spent many months learning the intricacies of desert warfare, made dozens of jumps from C130s, performed more than 200 helicopter abseils, and fired more light anti-tank weapons than he cares to remember.

Upon returning to academia, the author embarked upon a master's degree, supporting his studies by winning prize money in 100-km running races. After placing third in the World 100-km Championships and setting the North American 100-km record, Erik turned to ultra-distance triathlon, winning the World Endurance Triathlon Championships in 1995 and 1996. For good measure, he won the inaugural World Double Ironman Championships in 1995 and the infamous Decatriathlon – an event requiring athletes to swim 38 km, ride 1,800 km, and run 422 km. Non-stop!

In 1996, Erik pursued his Ph.D. at the German Space Agency. While conducting his studies, he won Ultraman Hawai'i and the European Ultraman Championships as well as completing the Race Across America bike race. Due to his success as the world's leading ultra-distance triathlete, Erik was featured in dozens of magazine and television interviews. In 1997, *GQ* magazine nominated him as the "Fittest Man in the World."

In 1999, Erik took a research job at Simon Fraser University. In 2005, the author worked as an astronaut training consultant for Bigelow Aerospace and wrote *Tourists in Space*, a training manual for spaceflight participants. In 2009, he was one of the final 30 candidates in the Canadian Space Agency's Astronaut Recruitment Campaign. Between 2008 and 2013, Erik served as director of Canada's manned centrifuge and hypobaric operations.

In addition to being a suborbital astronaut, triathlete, centrifuge operator and director, pilot, and author, Erik is an avid mountaineer and is currently pursuing his goal of climbing the Seven Summits. *Mars via the Moon* is his twentieth book. He currently works as an Assistant Professor in Commercial Space Operations at Embry-Riddle Aeronautical University where he spends time managing the suborbital spaceflight simulator and operationalizing pressure suits for use in space. He divides his time between the Space Coast, Sandefjord, Norway, and Kona, Hawai'i.

Acronyms

A-ETC	Architecture Et Cetera (A-ETC)
ALARA	As Low As Reasonably Possible
ALHAT	Autonomous Landing and Hazard-Avoidance Technology
ARED	Advanced Resistive Exercise Device
ARM	Asteroid Redirect Mission
ARS	Acute Radiation Sickness
ATHLETE	All-Terrain Hex-Limbed Extra-Terrestrial Explorer
BEAM	Bigelow Expandable Activity Module
BLSS	Bioregenerative Life-Support System
CAD	Computer-Aided Design
CAT	Computed Assisted Tomography
CATALYST	Lunar Cargo Transportation and Landing by Soft Touchdown
CLSE	Centre for Lunar Science and Exploration
CMO	Crew Medical Officer
COPUOS	Committee on the Peaceful Uses of Outer Space
CSF	Cerebrospinal Fluid
CSLA	Commercial Space Launch Act
CT	Computed Tomography
DGB	Disk Gap Band
EDL	Entry, Descent, and Landing
ESAS	Explorations Systems Architecture Study
FAA	Federal Aviation Administration
FPGA	Field Programmable Gate Array
GCR	Galactic Cosmic Radiation
GTO	Geosynchronous Transfer Orbit
HEOMD	Human and Exploration Operations Mission Directorate
HTP	Hypersonic Transition Problem
ICF	Inertial Confinement Fusion
ILD	International Lunar Decade

ILDWG	International Lunar Decade Working Group
ILOA	International Lunar Observatory Association
ILSWG	International Lunar Survey Working Group
ISECG	International Space Exploration Coordination Group
ISRU	In-Situ Resource Utilization
ISS	International Space Station
JFCC–Space	Joint Functional Component Command for Space
LEAG	Lunar Exploration Analysis Group
LEDA	Lunar European Demonstration Approach
LEO	Low Earth Orbit
LK	Lunniy Kabina
LLCS	Lunar Laser Communications System
LPS	Lunar Positioning System
LRO	Lunar Reconnaissance Orbiter
LTV	Lunar Transport Vehicle
MCF	Magnetic Confinement Fusion
MMAMA	Moon Mars Analog Mission Activities
MORO	Moon Orbiting Observatory
MS-FACS	Microwave Sinterator Freeform Additive Construction System
NAS	National Academy of Science
NCRP	National Council on Radiation Protection
NEO	Near Earth Object
NGL	Next Giant Leap
NORAD	North American Aerospace Defense Command
PGM	Platinum Group Metal
PISCES	Pacific International Space Center for Exploration Systems
POLO	Polar Orbiting Lunar Observatory
RAD	Radiation Assessment Detector
RESOLVE	Regolith and Environment Science and Oxygen and Lunar Volatiles Extraction
ROR	Rules of the Road
SDI	Strategic Defense Initiative
SEC	Shackleton Energy Company
SEPP	Solar Electric Primary Propulsion
SLS	Space Launch System
SPE	Solar Particle Event
SSPS	Space Solar Power System
STP	Supersonic Transition Problem
TDS	Terminal Descent System
TEI	Trans Earth Injection
TRN	Terrain Relative Navigation
VASIMR	Variable Specific Impulse Magnetoplasma Rocket
VIIP	Visual Impairment Intracranial Pressure
VSE	Vision of Space Exploration
WHO	World Health Organization

Kick-Starting the Next Giant Leap

"You need to tell your story better. You need a better story to tell."
James Cameron speaking to engineers working on the Constellation program, 2006

Momentum is building for a return to the Moon. NASA's international partners on the International Space Station are in favor of returning to the lunar surface, as are India and China. The National Research Council is too, stating: "Of the several pathways examined, the one that does not include a meaningful return to the Moon – that is, extended operations on the lunar surface – has higher development risk than other pathways." The horizon goal may be Mars, but the political, funding, and technological and medical infeasibility of such an objective means the next logical step is a return to the Moon. While much has been learned about the Moon over the years, we don't understand its resource wealth potential and the technologies to exploit those resources have yet to be developed, but there are a number of companies that are developing these capabilities. And, with the discovery of water in the lunar polar regions, plans are in the works to exploit these resources for fuel for transportation operations in *cis*-lunar space and in low Earth orbit (LEO).

"If God wanted man to become a space-faring species, he would have given man a moon."
Krafft Arnold Ehricke, rocket-propulsion engineer and advocate for space colonization

Ehricke was right! The time has come for commercial enterprise to lead the way back to the lunar surface. Embarking on such a venture requires little in the way of new technologies. We don't need to develop super-fast propulsion systems like those required to get us to Mars safely, nor do we need hundreds of billions of dollars that the experts reckon it will cost to transport humans to the Red Planet. No, what we do need is a place to test the technologies and deep-space experience that will enable us to build a pathway that will lead us to Mars. That place is the Moon and this book explains why.

This book begins with an assessment of the very real and very lethal "dragons" that mean a manned mission to Mars using chemical propulsion is a death sentence. Chapter 1 describes these mission killers in detail. Of all the physiological mission killers, radiation is the most dangerous dragon of all, since it not only increases the risk of cancer but also accelerates bone loss through a process of osteoradionecrosis. If that wasn't bad enough, deep-space radiation can also cause brain damage and cataracts. Compounding the effects of radiation are the risks of visual impairment, dramatic bone loss, and the wasting-away of muscles. Imagine a half-blind, brain-damaged, and severely weakened astronaut facing the challenge of the entry, descent, and landing (EDL) sequence. And, talking of EDL, this is the most lethal dragon, and one that may be the most difficult to tame. On reading Chapter 1, you may think that I'm not in favor of a manned mission to Mars. Nothing could be farther from the truth. I'm Norwegian and my country has a rich heritage of bold exploration. There is nothing I want more than for humans to set foot on Mars but it is absurd to think that reality will occur anytime in the next 20–30 years short of a series of technological breakthroughs. So Chapter 1 is a dose of realism.

In Chapter 2, we take a look at how humans may return to the Moon via a government-sponsored mission or series of missions. Russia has been making noises about landing on the Moon and establishing a base sometime in the late 2020s time frame while China has openly admitted that its next bold goal is the lunar surface. And China isn't interested in a "flags and footprints" mission: China wants to set up shop and begin mining helium-3. Another possible government-sponsored mission may be one led by the European Space Agency (ESA), which is developing many of the technologies that will be required to establish a permanent presence on the Moon. And then there's the United States. NASA's chief has stated many times that a manned mission to the Moon is not on the agenda any time soon, but NASA is still sending and planning missions to our nearest neighbor. Could a manned mission take the place of the much publicized asteroid-retrieval mission perhaps?

While a government-funded manned mission will head to the Moon eventually, chances are a commercial venture will be the first to return humans to the lunar surface. But what sort of markets exist on the Moon and are these markets financially viable? Chapter 3 adopts the Tauri strategy of examining possible markets such as tourism, resource extraction, technology, test and demonstration, and support and supplies. It then identifies the most likely commercial enterprises that are targeting a business on the Moon with particular attention focused on Shackleton Energy Company and Golden Spike.

Regardless of whether it is a government-funded mission or a commercial enterprise that returns humans to the Moon, any such venture must have access to the technology to realize such an objective; radiation-proofed habitats must be built; life-support systems capable of supporting extended stays must be developed; autonomous hazard-avoidance landing systems must be tested; and in-situ resource utilization processes must be proven in the field. Chapter 4 describes how these technologies are being developed and how they may be tested on the lunar surface.

Since this book argues that the Moon must serve as a test bed for an eventual manned mission to Mars, the human element is obviously integral to both ventures. Given the hazards of living in deep space, Moon- and Mars-bound astronauts will be subject to some new selection procedures, including pre-emptive surgery and sterilization. Chapter 5 describes how potential lunar crewmembers will undergo an appendectomy, gall-bladder removal, and sterilization before being genetically tested to ensure only those resistant to vision impairment and those with the highest bone density are selected. This chapter also describes the ultimate Mars analog mission in which a crew of four spend six months in lunar orbit to simulate a trip to the Red Planet, before landing on the Moon for a 12-month surface stay. The crew then blasts off back into lunar orbit for another six months before returning to Earth so that scientists can study them.

All the talk of mining helium-3 and establishing commercial operations on the lunar surface has inevitably led some to question the legality of such ventures. The truth is there is little in all the legalese that can prevent anyone from landing on the Moon and doing pretty much whatever they want. In an effort to provide some clarity on the issue of property rights on the Moon, Chapter 6 delves into the Moon Treaty and asks what might be done to establish some kind of legal framework for those working on the Moon.

Constellation. VentureStar. The Ukraine Crisis. The Chelyabinsk meteor. Headwinds or tailwinds? Chapter 7 describes how certain events may decelerate a return to the Moon and how some initiatives such as the International Lunar Decade may advance the goal of boot prints on the lunar surface. Finally, Chapter 8 discusses the business case. Commercial operations on the Moon with images of rovers scurrying around getting on with their business look convincing on a PowerPoint presentation, but just how viable is the process of establishing a commercial operation on the Moon? And how much will it cost? Is it reasonable to expect that investors will cough up the US$25 billion that Shackleton Energy Company is seeking to kick-start their propellant plant on the Moon?

1

Martian Dragons

Credit: NASA

Space agencies have been launching astronauts into space for decades, so you might assume scientists would have a handle on the physiological effects of spaceflight by now. The problem is they don't. But they do know there are many, *many* obstacles to overcome before we can even begin to think of sending crews on long-duration missions beyond Earth orbit using conventional chemical propulsion – radiation, vision impairment, brain damage, bone loss, to name just a few. And, even if these medical hurdles are resolved, any crew bound for Mars must still face the entry, descent, and landing (EDL) challenge. So Mars is very much a project for the next generation because we need to return to the Moon and use that as a test bed to resolve these problems which are described here.

> "It is not just the dose rate that is the problem; it is the number of days that one accumulates that dose that drives the total towards or beyond the career limits. Improved propulsion would really be the ticket if someone could make that work. The situation would be greatly improved if we could only get there quite a bit faster."
>
> Dr Cary Zeitlin, of the Southwest Research Institute,
> interviewed by the BBC in May 2013

DRAGON #1: RADIATION

Manned Mars mission proponents are fond of arguing that the medical risks of a multi-year journey to and from the Red Planet are exaggerated. The "Humans to Mars in 10 Years" and the "Mars Direct" crowds contend that, since astronauts spend six months on board the International Space Station (ISS), there is no big deal spending the same amount of time traveling to Mars. Well, first of all, a manned Mars mission will require crews to spend that six-month period outside low Earth orbit (LEO) in deep space. And, in deep space, things get ugly – *real* ugly, especially when we're talking about radiation (Figure 1.1). On board the ISS, astronauts are afforded the protection of Earth's magnetic field but, beyond Earth orbit, astronauts must face a deadly cocktail of galactic cosmic rays (GCRs) and solar particle events (SPEs). The biological damage GCRs and SPEs inflict means the risk to astronauts is way beyond unacceptable. And, to reduce that risk, it will be necessary to understand the deep-space environment and develop operational procedures to protect crews against that environment. How? Well, we can begin by sending robotic probes armed with radiation dosimeters. This is what NASA did when they dispatched the Curiosity rover (Figure 1.2) to Mars. During Curiosity's cruise phase in 2012, a Radiation Assessment Detector (RAD) measured GCR and SPE radiation, and the news was not encouraging. During its mission, Curiosity's RAD measured radiation that would have equated to a human receiving a whole-body computed axial tomography (CAT) scan every five days [1–3]. And that amount would increase the chances of an astronaut developing cancer by 5%, which exceeds NASA's threshold.

Based on information returned by robotic missions, space agencies can go to work and develop risk-projection models and calculate uncertainty assessments which can then be integrated into a future manned mission. Obviously, a crew will want to know there is a low level of uncertainty and they will also want to know what the thresholds for radiation exposure are, which is why NASA has developed Permissible Exposure Limits (PELs),

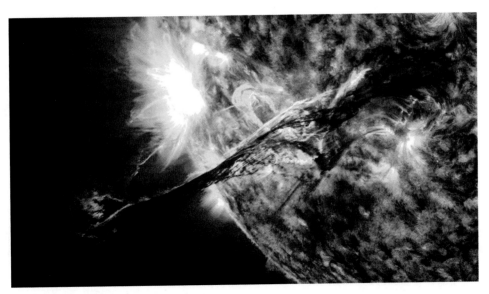

1.1 Solar flares spew out mission-killing radiation. Astronauts en route to Mars would require water walls 10 meters thick to provide them with the same protection as Earth's atmosphere. Until the radiation shielding problem is figured out, the sensible destination is the Moon. Credit: NASA

1.2 NASA's Curiosity rover, which carried a Radiation Assessment Detector (RAD) that was kept switched on throughout the Earth–Mars transit. The news was not good: astronauts making the same journey would increase their lifetime risk of cancer by 5% and that is beyond NASA's career limits. Deep space is full of lethal radiation and there is no escape, no matter how good the radiation shielding is. Imagine the following: you're standing in the middle of the desert when the wind picks up and all of a sudden you're sand-blasted by swirling particles. But, in deep space, radiation won't bounce off you like those sand particles. Radiation tears right through you, causing trails of potentially lethal damage behind. Credit: NASA which in turn are based on the National Council for Radiation Protection (NCRP) recommendations. NASA also supports the As Low As Reasonably Achievable (ALARA) principle when it comes to designing mission profiles but, while these acronyms sound comforting, there is significant uncertainty in predicting radiation exposure because scientists just don't have sufficient knowledge about deep-space radiation and its effect on the body. It's a field of research known as *radiobiology* which is basically the study of the effects of radiation on the human body [4]. The problem with radiobiology is that there are very, *very* few humans who have been subjected to heavy ions for a sustained duration. So what can scientists do to characterize the space-radiation environment for a Mars mission? Well, one of the first things they can do is determine acceptable levels of risk. Of interest, in 1967, in the Apollo era, the National Academy of Sciences (NAS) did not recommend radiation limits, arguing that limits might put the lunar missions at risk. That perspective changed in 1970 when the NAS Space Science Board outlined career radiation doses for astronauts, noting that the main risk was the increased probability of cancer as a result of increased radiation exposure while working in space. The NAS recommendations were used by NASA until 1989, by which time more epidemiological data had been acquired that allowed a more accurate dose limit to be calculated. But these were dose limits for LEO workers, not for those venturing beyond Earth orbit. Based on these data, the NCRP (Report No. 98) made the recommendation that astronauts should not be exposed to career radiation levels resulting in more than a 3% excess cancer risk [5]. Much of the data used to arrive at this number were derived from workers in nuclear facilities and these groups' risk was much lower than the 3% excess risk recommended for astronauts. So why were astronauts placed in a higher-risk category? Well, the NCRP was of the opinion that manned spaceflight held scientific benefits that justified the increased risk, although the council also noted that improvements had to be made to radiation protection for crews. Let's provide some occupational perspective and consider airline pilots for a moment (Figure 1.3). Pilots are not considered as radiation workers but they are exposed to between one and five millisieverts of radiation per year. An astronaut working on board the ISS? He or she will receive a dose of about 80 millisieverts per year. But how much more likely are those long-duration crews to get cancer? It's difficult to say because astronauts are an exceptionally healthy population group, so it is difficult to make occupation-against-occupation comparisons. Nevertheless, NASA has defined PELs for its crews and it is instructive to understand how these limits were determined if we are to appreciate just how dangerous traveling to Mars will be.

To assess radiation exposure and calculate the increased chance of fatal cancer, a series of factors must be considered. First, the body has different tissues and some tissues are more or less susceptible to radiation than others. Second, the absorbed dose must be considered because different types of radiation have different effects on the body. Next, the average risk to each tissue type must be calculated because, by doing this, it is possible to arrive at a summed radiation dose based on the weighting of the tissues and radiation type. Finally, mission duration, age, and gender must be inputted into the equation and this information scaled to mortality rate for radiation-induced death. An evaluation of cancer risk can then be calculated. Sounds like a very precise way of determining risk but there are all sorts of confounding variables. For example, this calculation doesn't factor in radiation sensitivity or the uncertainty of space-radiation dosimetry. But it's a start. So how

1.3 According to research published in the *Aviation Space Environmental Medicine* journal, an airline pilot averaging 673 hours per year will receive an average cosmic ray does of 2.27 mSv. Credit: NASA

does this apply to our manned Mars mission? Well, it's difficult to say because, as noted before, we just don't know much about deep-space radiation. But we do have information about certain GCR and SPE events, and we know there were major solar events in August 1972, October 1989, and July 2000. What would have happened if a crew had been bound for Mars during these events? Thanks to predictive modeling,[1] it is possible to estimate radiation doses, which is what a group of NASA scientists did for the 1972 SPE (Tables 1.1 and 1.2).

So what do these numbers mean? Well, the higher the numbers, the more severe the symptoms of acute radiation sickness (ARS). ARS progresses through three stages. The first is the Prodromal Stage, which is characterized by nausea and vomiting. These symptoms may last up to a few days. The Latent Stage follows the Prodromal Stage and is characterized by the patient looking reasonably healthy, although internally cells may be dying. At the Manifest Illness Stage, the symptoms are specific to the syndrome (Table 1.3).

One of the reasons SPEs are so dangerous is because they spew out huge radiation doses in a very short period of time. In radiation terminology, this is known as a *high dose-rate* and being exposed to a very high amount of radiation in a very short period of time inevitably leads to ARS symptoms (Table 1.3). Some of these symptoms may develop within minutes or hours, while delayed effects such as cataracts and cancer may develop over longer periods. And we must also consider the radiation from GCRs which comprise

Table 1.1 Dose equivalent and dose in critical body organs within an aluminum spacecraft during the August 1972 SPE (cGy) [1].

Organ	Spacesuit	Pressure vessel	Shelter
Skin	4,830	2,120	76
Lens	2,400	1,420	71
Blood forming organs	157	130	17

Table 1.2 Dose equivalent and dose in critical body organs within a polyethylene structure during the August 1972 SPE (cGy) [1].

Organ	Spacesuit	Pressure vessel	Shelter
Skin	3,620	1,540	40
Lens	2,080	1,150	38
Blood forming organs	151	120	10

Table 1.3 Acute radiation syndromes.

Syndrome	Dose (Gy)	Prodromal Stage	Latent Stage	Manifest Illness Stage
Bone marrow	>0.7	Nausea, vomiting; onset 1 hr to 2 days after exposure	Stem cells in bone die; stage lasts up to 6 weeks	Anorexia, fever, fall in blood cell count; death within months of exposure
Gastrointestinal	>10	Severe vomiting, nausea; onset within hours of exposure	Cells lining gastrointestinal tract dying; stage lasts < 1 week	Malaise, anorexia, fever, dehydration; death within 2 weeks of exposure
Cardiovascular/ central nervous system	>50	Confusion, severe nausea, loss of consciousness; onset within minutes of exposure	Partial functionality	Convulsions, coma; death within 3 days of exposure

heavy nuclei that tear through the body causing more damage. GCRs are particularly damaging because they destroy tissues at the molecular level. Another concern for mission planners is the very high penetration power of this type of radiation which means it is practically impossible to shield against [6].

So what are the risk limits? Truth is, we just don't know. Sure, we have the data from SPE events and the radiation levels measured by Curiosity, but determining risk limits is almost akin to witchcraft because so very little is known about deep-space radiation [7]. While scientists can model organ exposures for GCRs and SPEs for planetary surfaces and spacecraft constructed of different materials with a reasonable degree of accuracy, the

1.4 In the first part of the ultimate Mars mission analog, our intrepid volunteers would spend the first six months orbiting the Moon before landing and then spend another year or so conducting the same activities as they would on Mars. Then, after their surface increment, they would return to lunar orbit to spend another six months simulating the return journey to Earth. Credit: NASA

problem with these radiation events is that they are so variable and unpredictable (see sidebar). So what to do? Well, there is one option and that is to conduct an analog mission (Figure 1.4) that simulates a manned mission to Mars. The analog mission would go something like this. First, a launch into a lunar orbit, where the prospective Mars colonists would spend six months orbiting the Moon to simulate the length of a trip to Mars using chemical propulsion. Why orbit the Moon and not Earth? To simulate radiation exposure naturally. And, during the six months orbiting the Moon, the crew would have to take care of themselves in the event of an emergency, just like they would have to do if they were traveling to Mars. After six months on orbit, the crew would land on the Moon and spend a year on the surface simulating a Mars surface mission, after which they would launch back into lunar orbit and spend another six months simulating a return from Mars. All told, the crew would be exposed to deep-space radiation for three years. On return to Earth, those who survived would be subjected to myriad medical tests to determine just how much damage all that deep-space radiation inflicted. Unethical? Perhaps, but there are enough "Mars Today" fanatics out there that volunteers won't be hard to come by.

> *NASA's Exploration Systems Architecture Study (ESAS) Report*
>
> In 2004, in the days of the Vision for Space Exploration (VSE), NASA developed a plan to extend human surface operations on the Moon and Mars. This plan was outlined in the agency's Exploration Systems Architecture Study (ESAS) Report, which featured a section on radiation. NASA calculated that, even with 30 grams per centimeter of regolith shielding, the radiation emitted by a SPE during solar maximum would exceed an astronaut's 30-day limit on the lunar surface. As for Mars, the calculations weren't much more encouraging. The ESAS Report estimated the probability of the crew being hit by a 1972-category SPE was about 10% for the outbound and inbound legs of the trip, and that even 10 grams of aluminum shielding per centimeter wouldn't be sufficient to protect the crew.

Another danger from GCRs is the type of damage they cause [7, 8]. Cosmic rays are much, *much* more aggressive than other types of radiation, and one of the consequences of being exposed to this type of radiation is the development of tumors. Now, for those who have been unfortunate enough to have a loved one suffering from a cancerous tumor, you will know that treating such a condition requires intensive medical treatment – something that will most definitely not be available en route to the Red Planet. And, if the development of tumors wasn't bad enough, those cosmic rays also tear through genetic material (Figure 1.5), causing genetic mutations. We'll get back to the problems of dealing with

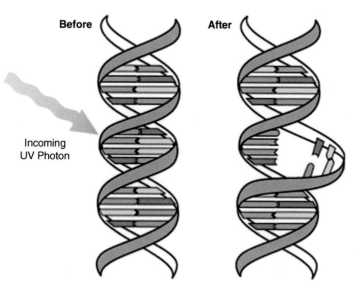

1.5 Radiation can tear apart genetic material with dire consequences: tumors, cataracts, increased cancer risk, sterilization, brain damage – the list goes on and on. Credit: NASA

tumors and genetic mutations on a spaceship shortly but, before we do, it's important to understand the degree of exposure Mars-bound astronauts will face. Based on modeling of GCRs and SPEs, scientists reckon it would take only 72 hours for each cell of each crewmember's body to be hit by a high-energy proton [9]. When you consider that your body contains trillions of cells, that adds up to an awful lot of hits! And, during a six-month trip to Mars, the chances are high that each astronaut would be hit by at least one heavy iron nuclei – the type that rip apart DNA and cause cancers to develop: mostly of the lung, breast, and colorectal variety [10]. And, if those cancers do develop, what then? Well, here on Earth, the standard treatment is chemotherapy, which involves injecting powerful medications into the body to destroy the cancer cells. This mode of therapy inevitably destroys the healthy cells with the result that chemotherapy causes painful side effects. Since the side effects are so debilitating, chemotherapy drugs are given intermittently, to give the non-cancerous cells time to heal. Typically, a course of chemotherapy may take three months. Imagine having to deal with this on a spaceship! But if one or more astronauts are diagnosed with tumors, chances are that surgery will be the only option. Cancer surgery may begin with the removal of lymph nodes to help the surgeon determine the extent of the cancer and the chances of the patient being cured. This procedure is then followed by a particular type of surgery: cryosurgery, electrosurgery, laser surgery, laparoscopic surgery, or natural orifice surgery. It is highly unlikely any of these types of surgery will be available on an interplanetary spacecraft, but perhaps the crew medical officer (CMO) will give it a shot. What then? Well, first of all, the CMO would have to perform a series of blood tests, urine tests, and imaging tests to determine surgical needs. Maybe a transfusion will be required? Who knows. Next, options for anesthesia will be decided, and the risks discussed with the health care team (the CMO!). These risks will include the certainty of a lot of pain, possibility of infection, possible loss of organ function, bleeding, blood clots, and altered bowel function. Imagine having to deal with this in a reduced-gravity environment with limited medical supplies. It just isn't going to happen. Another reason we need to head back to the Moon first!

Now let's get down to the business of surgery. And let's not forget that it is unlikely that the CMO will be a trained surgeon. How would you like someone who isn't even a physician, never mind a surgeon, performing surgery on you? And, even if the CMO has been practicing his or her skills every day – very unlikely given the myriad mission demands – how will they deal with such problems as intra-abdominal bleeding in microgravity? Okay, so we know radiation is bad – so bad that some scientists reckon a trip to Mars could shorten an astronaut's life expectancy by between 15 and 24 years. But can anything be done? *Anything*? Well, we can wait for a faster propulsion system to be developed. VASIMR (Variable Specific Impulse Magnetoplasma Rocket; Figure 1.6) comes to mind, but this won't placate the "Mars Now" crowd because they want to travel there today.

DRAGON #2: CATARACTS

Radiation doesn't just cause tumors and cancers; it also causes cataracts. That's because the ocular lens happens to be very sensitive to radiation. And with continuous radiation bombardment, the outcome will be cataracts, so now you have a crew suffering from

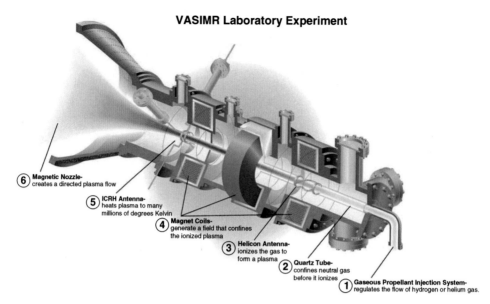

1.6 The Variable Specific Impulse Magnetoplasma Rocket (VASIMR) is our fast ticket to Mars. Using this propulsion system will reduce a six-month mission to just 39 days. In this rocket, gas is converted into plasma by ionization using a helicon coupler. In its cold plasma state, the gas is a mix of ions and electrons. A second coupler – the Ion Cyclotron Heating – pushes the ions that are then converted into thrust. You can read all about this revolutionary propulsion system in *To Mars and Beyond – Fast! How Plasma Propulsion Will Revolutionize Space Exploration* by Franklin Chang-Díaz and yours truly, published by Springer-Praxis. Credit: NASA and Ad Astra

cancerous tumors who also happen to be half blind. It's not a good mission outcome. And we're not speculating here. Research conducted by Cucinotta [11] analyzed the radiation-exposure dose of 295 astronauts, 48 of whom developed different types of cataracts. Of this group, 36 had been members of high-radiation Apollo missions (remember, these were short two-week flights, not a six-month Mars transit) and had developed cataracts as soon as four or five years after their mission [12]. How do cataracts happen? The answer lies in the lens of the eye (Figure 1.7), which must be crystal clear for you to have good vision. In a healthy eye, the lens focuses light onto the retina, thanks to the transparent fiber cells that lie in the center of the lens. But if the lens is damaged in some way, these fiber cells become clouded and it is this change in clarity that is termed a cataract. Here's a little more detail. Under normal terrestrial conditions, new fiber cells are created to replace old ones – a process that starts with the epithelial cells. Epithelial cells are a type of stem cell that cover the lens. When required, these cells flatten out and get rid of their nuclei, metamorphosing into transparent fiber cells. But, under radiation bombardment, this process doesn't work that well because radiation interferes with certain genes that control the life cycle of cells, including epithelial cells. The trouble with this interference is that it takes time for effects to be observed, which is a problem for space agencies

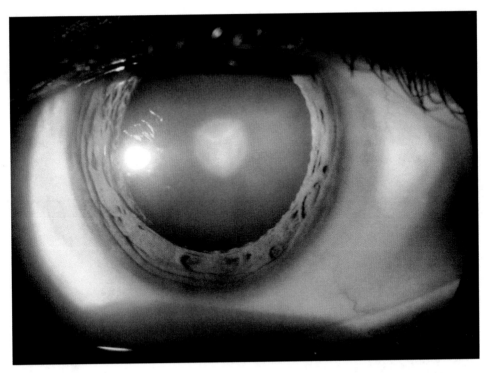

1.7 Cataracts – 36 former astronauts who took part in high-radiation missions (Apollo) have developed cataracts. These astronauts were exposed to deep-space radiation for no more than 12 days. Imagine the damage caused by a three-year mission – almost 100 times longer! Shielding, you say? Well, if you want the same protection as standing at 5,500 m altitude, you would need your spacecraft to be surrounded by a sphere with water walls five meters thick. Credit: Rakesh Ahuja, MD

worried about their astronauts developing cataracts because, by the time the first symptoms appear, it is too late to do anything about it. Another problem is the uncertainty about how radiation doses affect the onset of cataracts. For example, researchers don't know whether a high onset dose will accelerate the development of cataract and, if so, how high a dose is required.

The bottom line is that, by the time the lens is cloudy, it is too late to do anything about it. While techniques such as dynamic light scattering are being developed to detect early signs of damage to the lens (by measuring protein changes caused by radiation-induced oxidative stress), this hardly matters to Mars-bound astronauts because there are no effective countermeasures or treatments. Here on Earth, if a person is stricken with cataracts, the only effective treatment is surgery. Cataract surgery isn't a complex procedure, since it essentially involves replacing the cloudy lens with an artificial one. But even though a Mars-bound crew will be cross-trained in myriad medical procedures, it is highly unlikely any of them will be qualified to carry out cataract surgery. This type of surgery is performed using *phacoemulsification*, which requires the surgeon to make a small incision on the side of the cornea usually using a laser (imagine trying to do this in microgravity!). Once the incision is made, ultrasound is used to break apart the lens before vacuuming the fragments through the incision. It is a very precise procedure that requires a *very* skilled and *very* steady hand. Once the fragments have been cleaned out, the old lens is replaced with an artificial acrylic (or silicone) intraocular lens, which ultimately restores vision to normal. The patient than waits a few weeks before having the same procedure performed on the other eye. On Earth, the procedure is very safe but, as with any medical intervention, problems can arise. Eyes could become infected, the corneas could swell, retinas could become detached, or the implanted lens could become dislodged. Having said that, performing even this simple operation in a spaceship would be challenging at best: even in terrestrial hospitals, cataract surgery carries with it the risk of lens dislocation, retinal detachment, and infection.

DRAGON #3: BLINDNESS

"This is one that we don't yet have a good handle on, and it can be a showstopper for long-duration missions."

NASA scientist Mark Shelhamer

As if the possibility of suffering from cataracts wasn't bad enough, there is another eye problem that could be a mission killer. Some astronauts who spend several months living on the ISS develop vision problems. NASA's acronym for the condition is VIIP (Visual Impairment/Intracranial Pressure) and it has the agency concerned because no one really knows for sure why some astronauts suffer from the problem and others don't. VIIP is similar to papilledema, which is a medical term used to describe swelling of the optic disc caused by increased intracranial pressure (ICP). In most cases, astronauts regain their normal vision on return to Earth, but there are some astronauts for whom the effects linger for months. But on orbit, those astronauts afflicted by VIIP sometimes have to wear special reading glasses. While no astronaut has gone blind, if left unchecked, papilledema can ultimately lead to blindness, which is why NASA has assembled a VIIP task force to trouble-shoot the issue. Since the first VIIP summit in 2011, NASA has come up with a few theories that go some way to explaining the development and pathophysiology of the syndrome. One theory is that the build-up of cerebrospinal fluid (CSF) exerts pressure on the optic nerve, resulting in vision impairment, while another suggestion is that elevated levels of carbon dioxide on board the ISS may be a contributing factor. The problem is there is no smoking gun: symptoms include everything from choroidal folding to optic disc edema (Figure 1.8) to globe flattening and nerve kinking. Experts who have been investigating the problem suggest the reason there is no single factor is because there are so many inter-individual differences such as genetics, susceptibility, and differences in anatomical features. For example, some astronauts may have optic nerves that are more pliable under pressure, which means they would likely suffer more VIIP symptoms than an astronaut whose optic nerve is more resistant to pressure.

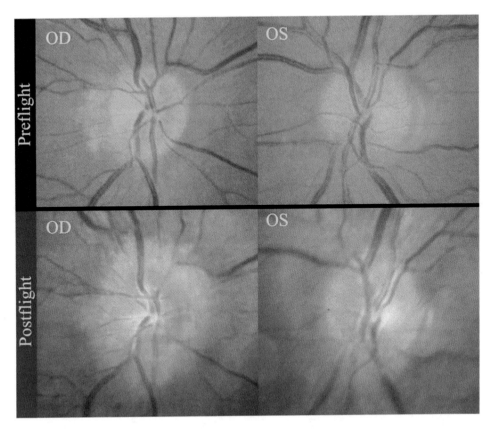

1.8 Optic disc edema, one of the many symptoms of the Visual Impairment Intracranial Pressure (VIIP) syndrome – one that NASA has no answer to. Imagine sending astronauts with visual deficits to Mars. Now imagine those astronauts also had cataracts and were half blind before they even landed! The VIIP problem is a big red flag for any long-duration mission beyond Earth orbit. Best to return to the Moon and buy the scientists some time to figure out a solution. Credit: NASA

DRAGON #4: BONE LOSS AND MUSCLE ATROPHY

"An artificial gravity system was deemed necessary for the MSM's outbound hab flight to (1) minimize bone loss and other effects of freefall; (2) reduce the shock of deceleration during Mars aerobraking, and (3) have optimal crew capabilities immediately upon Mars landing. Experience with astronauts and cosmonauts who spent many months on MIR suggests that if the crew is not provided with artificial gravity on the way to Mars, they will arrive on another planet physically weak. This is obviously not desirable. Unless a set of countermeasures that can reduce physiological degradation in microgravity to acceptable levels is developed, the only real alternatives to a vehicle that spins for artificial gravity are futuristic spacecraft that can accelerate (and then decelerate) fast enough to reach Mars in weeks, not months."

Excerpt from the Mars Society DRM report

The next Mars mission killers we'll discuss are bone demineralization and loss of muscle strength. Those of you who follow the long-duration missions on the ISS will no doubt be familiar with the post-landing scenes of astronauts and cosmonauts lying on couches waiting to be whisked away for post-mission rehabilitation (Figure 1.9). That's because, after six months on orbit, these crewmembers are in a significantly weakened state due to having lost an alarming amount of bone density and a worrying amount of muscle strength. This loss of bone density and muscle loss have been documented in hundreds and *hundreds* of research papers [13]: like radiation, these are sure-fire Mars mission killers and here's why. Astronauts lose bone 10–15 times faster than post-menopausal women and the scary thing is that no research has shown that all that bone is recovered after landing: there have been instances of some astronauts not having recovered their pre-mission bone density 10 years after landing (Thomas Reiter being one example) [14]. Ten years! And let's not forget, we're talking about astronauts who spent just six months on orbit. Extrapolate

1.9 Chris Hadfield relaxing after completing his stint as International Space Station (ISS) Commander. There will be no couches waiting for Mars crews! And no support personnel. And no rehabilitation team waiting to help the crew adjust to surface gravity. How do we know how deconditioned crews function in reduced gravity after six months in zero gravity? We don't, which is why it's a good idea to return to the Moon. Credit: NASA

Dragon #4: bone loss and muscle atrophy 15

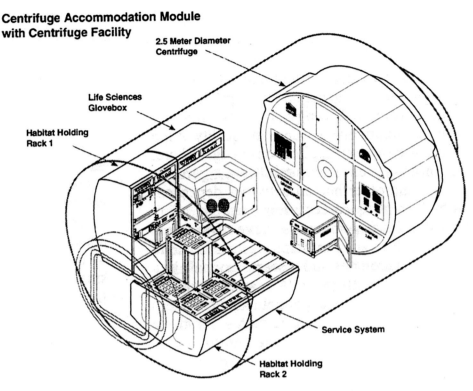

1.10 The Centrifuge Accommodation Module (CAM). A Mars-bound spacecraft can be provided with artificial gravity, the Mars crowd are fond of pointing out. I find this suggestion bizarre because we researchers know next to nothing about artificial gravity in space. How many astronauts have spent a week, a day, an hour even, being spun in artificial gravity in space in the last 20 years? Ten? Five? Answer: not one. How do we counteract the effects of disorientation and nausea when an artificial gravity is despun? How do we guard against the Coriolis cross-coupling effects? What is the optimal radius length? What is the correct gravity gradient to use? We don't know. Perhaps we would have had some of the answers if the CAM had been flown but it was parked in the car park at the Japan Aerospace Exploration Agency (JAXA) instead so we don't have the answers. Credit: NASA

the rates of bone loss observed during ISS missions to a multi-year Mars mission and you have an accident waiting to happen because astronauts wandering about the Red Planet would have bones that would be highly susceptible to fracture. The "Mars in a Decade" crowd contend that bone loss can be fixed with artificial gravity but it can't because we haven't proven the technology and we're still a long way from doing so. Perhaps if the Centrifuge Accommodation Module (Figure 1.10) had been flown on the ISS as originally planned (instead of rusting away in the parking lot at the Japan Aerospace Exploration Agency (JAXA)), we might be further along the path to solving the artificial-gravity problem, but we're not, so we're faced with the prospect of astronauts having the bone integrity of a 100-year-old woman wandering around on Mars.

1.11 Sunita Williams, during one of her many exercise sessions on board the ISS. Despite many astronauts training like marathoners (Williams "ran" the Boston marathon during her stint on orbit), the exercise programs still leave astronauts requiring rehabilitation on return to Earth. Credit: NASA

Exercise, you say? After all, on Earth, we know that exercise helps build bone density, maintains cardiorespiratory fitness, and builds muscle, but exercise isn't as effective in space. Even astronauts who trained like marathoners during their stint on the ISS (Figure 1.11) still returned to Earth with marked reductions in bone density and muscle strength. Let's take a look at some numbers from research performed by NASA to highlight exactly just how serious this problem is. In a recent long-term study, NASA researchers performed bone-density scans on 16 astronauts pre flight and one year post flight between Expeditions 2 and 8 and then compared these values against scans[1] performed on Earth-bound males and females. On average, astronauts lost 1.7% of outer bone mass and up to 2.5% inner bone mass per month during their long-duration missions [15]. Per month! Even after an entire year of recovery, during which time they had access to the very best of the best rehabilitation specialists (none of these on Mars remember!), these astronauts still had marked reductions in bone density, especially in the pelvis, hip, and long bones [13].

The story of muscle atrophy is little better. Mission after mission *after mission* has shown that astronauts lose muscle strength during long trips in space, despite exercising like demons during their flights. The loss of muscle strength is so profound that an

---

[1] Dual-energy absorptiometry and quantitative computed tomography were used.

astronaut in their 40s could end up with the strength of someone in their 80s after a six-month trip to Mars. What would happen in the event of an emergency? How would they complete routine tasks, never mind building habitats and conducting surface operations? And don't forget, some of these astronauts will be suffering from vision impairment, cataracts, and the early stages of cancer because we haven't figured out countermeasures for these yet! Despite exercise being a big part of an astronaut's day – crewmembers typically exercise for up three hours per day, six days a week – muscles still lose the ability to produce force. In an effort to mitigate the muscle-wasting problem, scientists have developed creative exercise devices such as the Advanced Resistance Exercise Device (ARED), which is being used by ISS astronauts today (Fitts), but there is still a long, *long* way to go before this mission killer is resolved.

DRAGON #5: EARLY-ONSET ALZHEIMER'S AND BRAIN DAMAGE

> "Galactic cosmic radiation poses a significant threat to future astronauts. The possibility that radiation exposure in space may give rise to health problems such as cancer has long been recognized. However, this study shows for the first time that exposure to radiation levels equivalent to a mission to Mars could produce cognitive problems and speed up changes in the brain that are associated with Alzheimer's disease."
>
> Dr. M. Kerry O'Banion, Department of Neurobiology and Anatomy, University of Rochester Medical Centre

Alzheimer's! In astronauts? Surely not. The problem is radiation again. In addition to increasing the risk of cancer and cataracts, it seems GCRs could cause the acceleration of Alzheimer's-type symptoms. To arrive at this conclusion, researchers studied mice, comparing two groups of the rodents: one group that wasn't exposed to radiation and a second group that was exposed to the type of radiation Mars-bound astronauts will be exposed to during a six-month trip to the Red Planet. After being exposed to radiation, the two groups of mice were subjected to memory and cognitive ability tests that required them to remember objects or locations. The group that had been exposed to radiation had a much higher failure rate than the group that hadn't been exposed, suggesting cognitive impairment.[2] When the researchers examined the brains of the mice that had been exposed to radiation, they found changes in vasculature and an increased amount of *beta amyloid*, which is a protein plaque that is one of the features of Alzheimer's. More shielding, you say? Remember, we're talking about high-mass, highly charged particles that slice through shielding like the proverbial hot knife through butter. Sure, it is possible to shield against these particles, but you would probably need to encase the spacecraft in a two-meter-thick concrete block.

Now imagine a crew suffering from early-onset alzheimer's. This crew would be prone to increased anxiety and bouts of depression in the early stages, but what if the symptoms worsened? Well, in the later stages, the crew would be prone to emotional distress, aggression, delusions, hallucinations, agitation, to name just a few of the classic symptoms. Now

---

[2] Lapses in attention were observed in 64% of the mice exposed to radiation and slower reaction times were observed in 27%.

18 Martian Dragons

the CMO would have to break out the antidepressants and anxiolytics in the hope of treating the afflicted crewmembers: Celexa and Prozac for those suffering from irritability; Ativan and Serax for those unusually anxious; and Clorazil and Haldol for those experiencing delusions. On Earth, those suffering from these sorts of symptoms are often assigned a caregiver but, on the surface of Mars, the crew would be left to fend for themselves. And let's not forget that some of this crew might already be suffering from vision impairment, cataracts, and early cancer symptoms.

DRAGON #6: ENTRY, DESCENT, AND LANDING

EDL is perhaps the toughest of all the challenges to a manned mission to Mars (Figure 1.12). While there have been a number of successful unmanned missions, these flights have typically landed payloads that weigh only a few hundred kilograms whereas a manned mission

1.12 A robotic entry, descent, and landing (EDL) architecture. A manned EDL will not be so simple. Far from it. This is the most dangerous "dragon" of them all and with good reason. The "Mars Today" crowd say aerocapture can be used to land large payloads on the surface of Mars. Really! And how many times has this been achieved? And how large was the payload? Two tonnes? Fifteen? Using aerocapture or any of the myriad unproven EDL technologies proposed by the "Mars Today" crowd is pure fantasy. The EDL dragon is the biggest flaw of all the manned mission to Mars architectures. Credit: NASA will require the landing of a payload that weighs at least 20 tonnes. Let's imagine a manned mission to get an idea of the challenges involved. We'll being by inserting the manned spacecraft into Martian orbit using aerocapture (Figure 1.13). Once aerocapture (see sidebar) has been achieved, the vehicle must position itself into a parking orbit by executing propulsive maneuvers. From there, the vehicle de-orbits and decelerates using a combination of a lifting entry and parachutes before landing by means of propulsive descent. Sounds simple doesn't it? In reality, it is anything but, although the "Mars Tomorrow" crowd would like you to believe otherwise. Let's consider some of the constraints. First of all, there is the condition of the crew to consider. Remember, this crew has been in deep space for around six months and has been subjected to radiation exposure and prolonged weightlessness. Some of them may be half blind to boot. We know they will be in weakened state due to the effects of deconditioning and chances are that some of them may be suffering from radiation sickness. In this weakened state, the crew won't be able to tolerate re-entry decelerations higher than four or five Gs, and they will only be able to endure this for short durations. Ideally, the loading should be through the chest because humans can best tolerate the highest Gs in this orientation. But, when the vehicle transitions from entry to powered descent to landing configuration, it may be necessary to re-orient the crew. This not only causes problems when designing the interior of the vehicle, but may also cause the crew to be disoriented. At such a super critical phase of the flight, this is the last thing the crew or mission planners want.

Aerocapture

Aerocapture is used to decelerate interplanetary spacecraft by flying the vehicle through a planet's atmosphere as depicted in Figure 1.13. A vehicle preparing to land on Mars will be flying through the atmosphere at about six kilometers per second, so we're talking about very high velocities here. To bleed off all that speed, there are a number of options. One is to use a blunt-body design that uses a heat shield. This is similar to the design employed by the Apollo Command Module. Another option is using a ballute, which is half balloon, half parachute, but this system has yet to be proven. A third option is supersonic retropropulsion, but let's consider option number one to begin with. A vehicle speeding along at six kilometers per second will create an awful lot of heat and a single heat shield probably won't be sufficient to dissipate the heat and keep the crew cool. Two heat shields, you say? Well, that's a heavy mass penalty which reduces the payload delivered to the surface. Ballutes, then? In theory, these might work but there is a long way to go before this system is proven since engineers must still work out how to ensure stability at very high speeds and also how to ensure a precision trajectory.

All this causes a problem for mission planners designing the entry corridor because the steeper the flight path angle, the higher the acceleration. Sure, you can make the flight path angle shallower but, if you do this, you risk an overshoot trajectory which will result in the vehicle skipping out of the atmosphere, and that would be a real shame after having

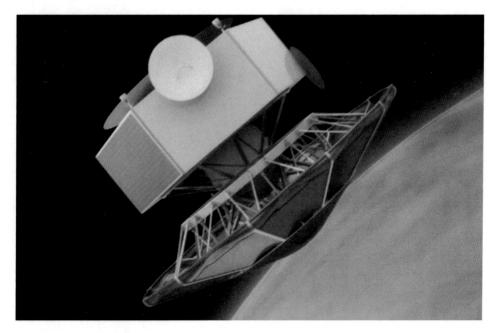

1.13 Aerocapture. Good in theory but a long, long way from being proven with a payload of the size needed to deliver a crew of four to the surface of Mars. A long, long way. Credit: NASA

spent six months traveling such a long distance. The challenge facing the EDL specialists is to find a navigable entry corridor that doesn't impose excessive deceleration on the crew and that is a tough ask. Let's take a look at the sort of challenges presented by EDL from start to finish.

We'll start by imagining a vehicle bulleting towards the Martian atmosphere about to begin the atmospheric entry phase. During its hypersonic entry, mission planners must ensure that the payload – the astronauts – are kept safe from the thermal loads and that the deceleration forces don't injure or kill the deconditioned crew. The mission planners must also ensure that the vehicle is delivered to an entry point at precisely the correct velocity and angle over the horizon. Too steep and the crew will be fried and knocked out by excessive deceleration. Too shallow and the vehicle will skip off the atmosphere and that will be the end of the mission. Another important requirement is ensuring all that kinetic energy is bled off and that the crew can look forward to a soft landing. Engineers are banking on the Martian atmosphere bleeding off much of that hypersonic speed but, even if it can slow the vehicle, there still needs to be some way to steer the spacecraft to a precise landing. When mission planners plan for a landing on Mars, they often refer to the landing ellipse. The size of the landing ellipse is determined by factors such as the navigational capabilities of the vehicle, the weather, and the ability of the vehicle to steer during the EDL phase. A good example to illustrate the concept of the landing ellipse is the Apollo Command Module (CM), which had to steer during its hypersonic re-entry. Thanks to its navigation system, the CM was guided to a precise landing very close to the recovery ships. The same

1.14 A disk-gap parachute. Sounds good in theory. Now let's see it tested on Mars. Credit: NASA

principle was used by the Lunar Excursion Module and the Space Shuttle. Now the chances are that the first manned spacecraft to attempt a landing on Mars will be of the capsule variety, so you may be wondering how such a vehicle can steer. Well, capsule spacecraft can fly thanks to the use of reaction control thrusters which are used to bank left and right. This, combined with retropropulsion, may be the way a future manned mission attempts a landing. But before attempting a landing, the vehicle must bleed off all that speed.

Many Mars EDL models use disk-gap band (DGB) parachutes (Figure 1.14) because DGBs tend to perform well at supersonic speeds in a low-density atmosphere such as that on Mars [16]. For an EDL plan that utilizes a DGB, the 30-meter-diameter parachute would be deployed at Mach 3 but, even with the deceleration capability of such a system, the vehicle will still need an additional means of bleeding off speed. And this is where it gets interesting because the vehicle doesn't just need to decelerate; it must also perform cross-range maneuvering, which requires acceleration and deceleration, and it also must be able to perform a search (in hover mode) and ultimately land. Each of these capabilities requires propulsive maneuvering, which requires engines, and these engines must be covered by the heat shield until the entry maneuvers begin, which in itself creates an engineering headache. Compounding this challenge is the decision to go with one heat shield (Figure 1.15) for aerocapture and another for the entry maneuvers, or to just use one heat shield for both. Let's assume only one heat shield is used. The next challenge is fabrication because the heat shield must be designed to be removed to expose the descent engines

1.15 NASA's Hypersonic Inflatable Aerodynamic Decelerator (HIAD). Credit: NASA

once entry maneuvers begin. This is a significant engineering challenge and here's why. First, there are two ways to drop the heat shield. One way is to drop it before firing up the engines and the other way is to open doors in the heat shield to expose the engines. The problem with the first option is the ballistic coefficient of the vehicle is greater than the ballistic coefficient of the heat shield and that causes a problem because, until the ballistic coefficient of the heat shield is greater than the vehicle's, that heat shield will stay forced against the bottom of the spacecraft. Why is this a problem? Well, if the heat shield detaches from the vehicle under these conditions, there is a chance the heat shield could ricochet off the vehicle [17]. So why not go with doors? The problem with doors is that incorporating these into a heat shield requires adding seams and penetrations and this reduces the effectiveness of the heat shield. How about positioning the engines on the nose of the spacecraft, then? Not likely, because this would require rotating the vehicle 180° and that would be very disorienting to our already discombobulated crew. Of course, all this deliberation about where to position the engines may be a moot point because we have to consider the mass of the heat shield together with all the other mission components such as life support, power, and fuel. In short, if the heat shield becomes too heavy, then we need a bigger parachute and, if we need a bigger parachute, then the chances are we can't safely land the crew [17].

But let's assume the hypersonic transition problem (HTP) can be solved. The spacecraft is traveling along at about Mach 5, the parachutes have been deployed, and the vehicle is now ready to begin the descent phase. Mach 5 is still a fair rate of knots and, while the Martian atmosphere will help to kill some of that velocity, the spacecraft will need to use a Terminal Descent System (TDS) to ensure the speed is manageable before preparing for the landing. This may prove difficult for a vehicle that weighs 20 tonnes or more because the vehicle will still be moving supersonically very close to the surface. This is known as the supersonic transition problem (STP) and at the core of this challenge is the velocity gap that exists below Mach 5: there just isn't a supersonic decelerator system that can slow a large payload quickly enough. In short, with the current EDL technology, a 20-tonne payload bulleting through the Martian atmosphere has just 90 seconds to decelerate from Mach 5 to Mach 1, re-orient itself from spacecraft to lander mode, deploy parachutes to bleed off the remaining speed, look for a landing site, avoid hazards, hover to the landing site, and touch down. Airbags can't be used because we've pretty much reached the limit of airbag technology and, even if we did use airbags, the deceleration forces would approach 20 Gs. That's okay for robots but not so good for a deconditioned and discombobulated crew. Parachutes are no good for the initial deceleration from Mach 5 because they can only open at speeds of around Mach 2.8. How about a lift vehicle like the Shuttle, then? Good idea, but the vehicle would have to be traveling very, very low to get the benefit of the atmosphere so, by the time the spacecraft had slowed to Mach 2.8, it would be too close to the surface. And even if the vehicle did have time to deploy a parachute, the canopy would need to be 100 meters in diameter and there is no known way of deploying a parachute that big. Thrusters then? Surely these would work. Yes, they would, but the fuel cost would be huge (as much as six times the mass of the payload!) to ensure the safe landing of a 20-tonne payload. But let's assume we could take along enough fuel. What then? Well, then there is the challenge of dealing with rocket plumes which happen to be extremely unstable, unpredictable, and dynamic. Imagine the vehicle streaking nose first through the Martian atmosphere at supersonic speeds with rocket plumes (Figure 1.16) moving around the vehicle. And these chaotic rocket plumes are moving around the spacecraft against extremely high dynamic pressure. The result would likely be loss of the vehicle from the twisting forces imparted by the plumes.

Once again, let's imagine the STP can be resolved. Now the spacecraft is zipping along at subsonic speed and can switch over to propulsion to adjust its trajectory in preparation for landing. Everything looks great, right? Wrong. Even with the STP resolved, the vehicle is just 1,000 meters above the surface and that doesn't leave much latitude for maneuvering. As the vehicle slowly sinks lower, the commander spots a viable landing site, but its six kilometers away and the wind is kicking up. Getting there is going to eat up an awful lot of fuel which the vehicle probably doesn't have. But it's not all doom and gloom

24 **Martian Dragons**

1.16 An inflatable heat-shield concept. Credit: NASA

because there are a few ideas out there that may make it possible to get a crew to Mars in one piece. One such idea is the inflatable supersonic decelerator, aka the Hypercone. This donut-shaped device would inflate around the vehicle at about 10 kilometers above the surface while the spacecraft is zipping along at Mach 4. Acting as an aerodynamic anchor, the Hypercone would slow the vehicle down to Mach 1. Problem solved. Not quite. But this Hypercone (Figure 1.17) would have to be up to 40 meters in diameter and structures that large are terribly difficult to control. And even if the Hypercone delivers and manages to decelerate the vehicle to sub-Mach 1 speeds, what then? Parachutes? Not likely, because these take time not only to deploy, but also to shed, which won't leave enough time to switch to propulsive systems. And the reason there is so little time is because that spacecraft is plummeting 10 times faster than a spacecraft would on Earth for the simple reason that the density of the Red Planet's atmosphere is just 1% of Earth's. But let's continue imagining and assume the first three phases – entry approach, hypersonic entry, and parachute descent – can be resolved. Now we're setting up for the landing.

As the vehicle descends to the surface, it will use radiometric data from other spacecraft and optical observations to find the landing target. Its sounds very precise but, even with all this information available, chances are the vehicle will be off course once it decelerates below supersonic speed due to the effect of local winds on the parachute. And let's not forget that we know very little about the winds below 10 kilometers altitude on Mars, so the vehicle may be blown way off course. Having said that, the vehicle will most likely weigh 20 tonnes or more, which means the surface winds may not have as much effect as on a robotic lander. Still, it's probably safe to assume the vehicle will be displaced from its target. But, thanks to the use of star trackers, an altimeter, Terrain Relative Navigation (TRN), and a velocimeter, a precision landing isn't out of the question – as

1.17 A Hypercone. Credit: NASA/Vorticity

long as the crew can avoid the slopes, craters, rocks, and other natural hazards, and as long as the crew have survived the supersonic retropropulsion entry loads, and as long as the vehicle has enough propellant to fly out the uncertainties as the spacecraft approaches the landing site. How much divert capability will the lander have? One kilometer? Five?

The bottom line in this EDL discussion is that 60% of all Mars missions have failed. The anxiety evoked by the myriad challenges presented by the task of landing *anything* on Mars has spawned phrases such as Six Minutes of Terror. We're a long, long way from developing a reliable EDL system. Best go to the Moon first!

REFERENCES

1. Townsend, L.W. Implications of the Space Radiation Environment for Human Exploration in Deep Space. *Radiation Protection Dosimetry*, **115**, 44 (2005).
2. Hellweg, C.E.; Baumstark-Khan, C. Getting Ready for the Manned Mission to Mars: The Astronauts' Risk from Space Radiation. *Naturwissenschaften*, **94**, 517 (2007).
3. Badhwar, G.D.; Cucinotta, F.A.; O'Neill, P.M. An Analysis of Interplanetary Space Radiation Exposure for Various Solar Cycles. *Radiation Research*, **138**, 201–208 (1994).
4. Barcellos-Hoff, M.H.; Park, C.; Wright, E.G. Radiation and the Microenvironment: Tumorigenesis and Therapy. *Nature Reviews Cancer*, **5**, 867–873 (2005).
5. Cucinotta, F.A.; Schimmerling, W.; Wilson, J.W.; Peterson, L.E.; Saganti, P.; Badhwar, G.D.; Dicello, J.F. Space Radiation Cancer Risks and Uncertainties for Mars Missions. *Radiation Research*, **156**, 682–688 (2001).
6. Cucinotta, F.A.; Kim, M.Y.; Ren, L. Evaluating Shielding Effectiveness for Reducing Space Radiation Cancer Risks. *Radiation Measurements*, **41**, 1173–1185 (2006).
7. Cucinotta, F.A.; Durante, M. Cancer Risk from Exposure to Galactic Cosmic Rays: Implications for Space Exploration by Human Beings. *The Lancet Oncology*, **7**, 431–435 (2006).
8. ICRP Publication 60, *Recommendations of the International Commission on Radiological Protection*, Pergamon Press Inc. (1991).
9. Kim, M.Y.; George, K.A.; Cucinotta, F.A. Evaluation of Skin Cancer Risks from Lunar and Mars Missions. *Advances in Space Research*, **37**, 1798–1803 (2006).

10. Ainsbury, E.A.; Bouffler, S.D.; Dorr, W.; Graw, J; Muirhead, C.R.; et al. Radiation Cataractogenesis: A Review of Recent Studies. *Radiation Research*, **172**, 1–9 (2009).
11. Cucinotta, F.A.; Manuel, F.K.; Jones, J.; Iszard, G.; Murrey, J.; et al. Space Radiation and Cataracts in Astronauts. *Radiation Research*, **156**, 460–466 (2001).
12. Chylack Jr., L.T.; Peterson, L.E.; Feiveson, A.H.; Wear, M.L.; Manuel, F.K.; et al. NASA Study of Cataract in Astronauts (NASCA). Report 1: Cross-Sectional Study of the Relationship of Exposure to Space Radiation and Risk of Lens Opacity. *Radiation Research*, **172**, 10–20 (2009).
13. Lang, T.F.; LeBlanc, A.D.; LeBlanc, A.D.; Evans, H.J.; Lu, Y.; Genant, H.K.; Yu, A. Cortical and Trabecular Bone Mineral Loss from the Spine and Hip in Long-Duration Spaceflight. *Journal of Bone and Mineral Research*, **19**(6), 1006–1012 (2004).
14. Sibonga, J.D.; Evans, H.J.; Sung, H.; Spector, E.R.; Lang, T.F.; Oganov, V.S.; Bakulin, A.V.; Shackelford, L.C.; LeBlanc, A.D.; LeBlanc, A.D. Recovery of Spaceflight-Induced Bone Loss: Bone Mineral Density after Long-Duration Mission as Fitted with an Exponential Function. *Bone*, **41**(6), 973–978 (2007).
15. Carpenter, R.D., LeBlanc, A.D.; LeBlanc, A.D.; Evans, H.J.; Sibonga, J.D.; Lang, T.F. Long-Term Changes in the Density and Structure of the Human Hip and Spine after Long-Duration Spaceflight. *Acta Astronautica*, **67**(1–2), 71–81 (2010).
16. Lafleur, J.; Verges, A. Parachutes and Propulsive Descent for Human Mars Exploration, AIAA Student Conference Paper, Starkville, 3–4 April 2006.
17. Wells, G.; Lafleur, J.; Verges, A.; et al. Entry, Descent and Landing Challenges of Human Mars Exploration, AAS 06-072, Breckenridge, CO, 4–8 February 2006.

2

Government Anchors

"The United States will not be sending a human to the Moon anytime soon because we can only do so many things. I don't know how to say it any more plainly. NASA does not have a human lunar mission in its portfolio – and we are not planning for one."
NASA Chief Charlie Bolden at a meeting of the Space Studies Board and the Aeronautics and Space Engineering Board

"NASA is not going to the Moon with a human as a primary project probably in my lifetime."
Charlie Bolden quoted in SpacePolitics.com blog

Credit: NASA

It was President Nixon who cancelled the Apollo Program. He made sure he delayed the announcement until he had been re-elected and, since that bleak moment in history, the Moon has been off the radar and it still is as far as NASA-sponsored missions are concerned if Charlie Bolden is to be believed. In 2004 there was half-hearted talk of a return to the Moon by way of President George W. Bush's initiative to have American astronauts walking on the surface by 2020, but nothing came of that thanks to President Obama. In 2015, NASA's manned exploration plan comprises a vague vision of visiting an asteroid with equally vague talk of a Mars mission in the very, *very* distant future. Meanwhile, China's Jade Rabbit is scooting across the lunar surface, after having landed on the Moon in 2013, and India and Japan are voicing interest in visiting Earth's closest neighbour. So which government agency will be first back to the Moon? Russia perhaps? China? The European Space Agency (ESA)? And when will NASA once again send its astronauts back there?

LUNAR NATIONS: RUSSIA

In February 2015, Russia announced it would continue working with the International Space Station (ISS) coalition until 2024, at which time it will begin preparing for manned missions to the Moon. In between abandoning the ISS and its first lunar missions, Russia plans to orbit its own space station to re-establish a permanent presence in space. If Russia's lunar plans are realized, it will continue a rich tradition in lunar exploration, since the Russians were the first to launch a probe (Luna 2) to the Moon all the way back in 1959 [1]. Luna 2 was followed by Luna 3, which photographed the far side of the Moon, and the Russians, emboldened by their success, planned a manned mission. Russia's manned Moon mission was devised by Sergei Korolev, who was an expert when it came to designing rockets and spacecraft. Unfortunately, Korolev was overshadowed by missile designer, Vladimir Chelomei, who employed Khrushchev's son – a shrewd political move that guaranteed Chelomei a big budget to develop his own rockets [2]. Set against this rivalry was the antagonism between Korolev and rocket engine designer, Valentin Glushko. Arguments between Korolev and Glushko in the 1930s had resulted in Korolev being sent to a labor camp, and the animosity that this event created hampered the progress of rocket engine development into the 1960s: Korolev was in favor of using cryogenic fuels whereas Glushko preferred engines fueled by hypergolic chemicals. It was just one example of the lack of coordination in the Russian space program. In 1962, Chelomei was assigned the task of developing a manned spacecraft capable of flying around the Moon. While Chelomei got on with this task, Korolev got busy developing the N1 rocket (Figure 2.1) which was supposed to launch a 40-tonne payload into low Earth orbit (LEO). But, since Korolev had fallen out with Glushko, he had no source for rocket engines. He eventually teamed up with Nikolai Kuznetsov, a designer of aircraft engines, who developed a conventionally fueled rocket engine for Korolev. Unfortunately, this engine was so weak that 30 of them were required on the first stage for the hypothetical lunar mission that Korolev envisaged.

In addition to being busy with the N1, Korolev was developing the multipurpose Soyuz spacecraft. Capable of docking and rendezvous, the Soyuz was designed to carry a crew of

2.1 The Soviet Union's N1 rocket that was built to ferry cosmonauts to the Moon. Credit: NASA.
Source: http://grin.hq.nasa.gov/ABSTRACTS/GPN-2002-000188.html

three. Variations of the vehicle, such as the Soyuz-A, were to be fitted with a translunar injection stage in Korolev's plans, but the Soviet leaders decided to go ahead with Chelomei's plans. But Korolev didn't give up his plans to land cosmonauts on the Moon. He lobbied for a manned circumlunar mission and in 1964 the Central Committee approved a plan to send one cosmonaut to the Moon in 1967/68. This flight would be preceded by a circumlunar mission manned by two cosmonauts. But, despite these plans, the Soviet lunar effort was still uncoordinated because in 1964 there were three design bureaus planning manned lunar landings. Ultimately, a decision in late 1964 decided to go ahead with a lunar mission that used the N1 rocket. Since the N1 required 26 engines to be fitted to the first stage, there was some consternation about the reliability of the system, but no test stand was ever built, and the Soviet Moon plans continued [3]. In 1965, Alexei Leonov performed the first spacewalk while wearing a prototype spacesuit. Buoyed by Leonov's success, mission planners knuckled down to devising a plan to land the first cosmonaut on the Moon in 1968. To that end, 28 new cosmonauts joined the ranks of the Soviet manned space program in October 1965 to fly the Soyuz.

The plan to actually land on the Moon centered on two spacecraft. One, the Lunniy Orbitalniy Krabl, or LOK, was a two-person lunar orbiting vehicle that was designated as the mothership during the flight to lunar orbit. Once in orbit around the Moon, a sole cosmonaut would exit the LOK and enter the Lunniy Kabina (LK – lunar cabin), which would then land on the Moon (Figure 2.2). After planting the flag and posing for pictures for posterity, the first Soviet on the Moon would climb back into the lunar cabin and join his comrade on board the LOK. The LK would be discarded, the cosmonauts would fire the trans-Earth injection (TEI) burn, and they would arrive home three days later. Mission accomplished! While the plan to actually land on the Moon was pretty clear-cut, the circumlunar mission was much less defined. The circumlunar mission was needed to test the systems and launchers that would be used during the manned mission but the problem was that the launch vehicles would not be ready in time. At about the same time as the Soviets were dealing with the delays in the development of the N1, there were also plans to send a robotic mission carrying a rover. One of the rover's jobs would be to carry landing beacons to be used by the LK vehicle during the descent. Its other role was to serve in a backup capability in the event that the LK was damaged during the descent. If this occurred, the rover would transport the cosmonaut to a backup vehicle.

Despite the vague mission architectures and the delays in developing the N1, the Soviet lunar program seemed to gather pace in late 1965. Unfortunately, with the death of Korolev in 1966, the program was robbed of its driving force and whatever momentum the program had slowed. Later that year, the N1 was reviewed and found to be lacking power, which forced yet another costly redesign that resulted in four engines being added to the 26. Meanwhile, the Americans were making inroads on the Soviets as the development of the Saturn V and Apollo spacecraft accelerated. The Soviets scrambled to try and catch up, but nothing seemed to work. The third Voshkod flight was put on hold for two months before being cancelled and then the whole program was terminated in an effort to prepare for the Soyuz, which was the centerpiece of the Soviet program. While it was a versatile spacecraft, the first three unmanned flights failed. With the third failure in February 1967, the Soviets found themselves way behind the Americans, which led to an impatient Brezhnev demanding a manned flight sooner rather than later. So, despite more than 200 faults having been identified in the Soyuz vehicle, Vladimir Komarov (Figure 2.3) launched on

2.2 The Russian lunar lander – the Lunniy Kabina (LK) – was capable of carrying just one cosmonaut. Credit: Eberhard Marx

Soyuz 1 on 15 April 1967. After just one day on orbit, following a flight plagued by problems, the veteran cosmonaut was ordered back. Unfortunately, due to a parachute malfunction, the Soyuz crashed and Komarov was killed.

Undeterred, the Soviets pressed on with the L1 circumlunar mission. The Soviets had planned to fly four unmanned flights in early 1967 before flying a manned circumlunar

2.3 Vladimir Komarov was selected to command the Soyuz 1 as part of the Soviet Union's bid to land cosmonauts on the Moon. Komarov's flight in April 1967 made him the first cosmonaut to fly more than once. He also became the first to die during a space mission when his capsule crashed following a parachute malfunction. Credit: AFP

flight in June 1967. But, due to the Komarov tragedy, the unmanned flights were delayed until September and November 1967. Neither flight was successful but the Soviets didn't abandon their N1 plans. On 15 July 1968, an attempt was made to launch the N1 but this resulted in another tragedy when an over-pressurized oxidizer tank exploded, killing three workers. Meanwhile, NASA had recovered from the fire that killed Grissom, White, and Chaffee, and were on course to attempt manned a flight to lunar orbit in December 1968. The Soviets responded by launching the unmanned Zond-5 in September 1968. While the Zond-5 became the first spacecraft to fly around the Moon, it was a minor first compared to the planned Apollo 8 mission. A day after the launch of Apollo 7 on 11 October 1968, the Soviets launched Soyuz 3 with cosmonaut Georgi Beregovoi on board. Next was the Zond-6 flight which was planned for November. The vehicle, carrying a biological payload, was launched successfully on 10 November. The spacecraft flew past the Moon but depressurized a few hours before re-entry, killing all the animals. Despite the failure, the Soviet propaganda machine announced the flight had been successful, but plans for the launch of the L1 were coming to an end. With the success of Apollo 8, the Soviet L1 lunar program began scaling down. A manned circumlunar flight was considered for April 1970 to celebrate Lenin's birthday but this was not approved [4].

With the death of the L1 program, the Soviets still had a theoretical chance of matching the Americans with their N1 rocket [5, 6]. The cosmonauts had started their lunar training in Star City in March 1968 and the commander for the first manned mission had been chosen, the honor having been given to Alexei Leonov. But before the lunar mission, the Soviets had to practice dockings and spacewalks. To that end, on 14 January 1969, Soyuz 4 launched, carrying Vladimir Shalatov. This flight was followed the next day by Soyuz 5, with Boris Volynov, Alexei Yeliseyev, and Yevgeni Khrunov on board. After the two vehicles had docked, Khrunov and Yeliseyev donned spacesuits and conducted a spacewalk to Soyuz 4. Confident in their docking and spacewalk procedures, the Soviets were ready to test their family of lunar-landing vehicles. First to be launched was the lunar rover on 19 February 1969. The rover's life was short-lived when the booster exploded less than a minute after launch. Undeterred, the Soviets readied the first N1 launcher which rose skyward on 21 February 1969. But just 66 seconds into flight, a fire started at the end of the first stage and range safety officers destroyed the vehicle. A month after the N1 failure, Apollo 9 had tested the lunar module and, in May 1969, Apollo 10 closed to within 15 kilometers of the lunar surface. Time was running out for the Soviets. Another attempt was made to launch the N1 on 3 July 1969 but, less than 10 seconds after launch, debris found its way into the oxidizer pump of one of the engines. The engine exploded and the thrust coordination system commanded the remaining engines to shut down. The booster fell back onto the pad and the resulting conflagration almost destroyed the pad. A few weeks later, Neil Armstrong walked on the Moon. The Soviets had been well and truly beaten. And humiliated. There was talk of extended stays on the Moon and, with that in mind, the Soviets began sending unmanned Luna vehicles to the lunar surface, including Luna 16 which returned some soil. Luna 16 was followed by the visit of a lunar rover in October 1970. Meanwhile, development of the advanced N1 program continued with the objective of landing on the Moon. A N1 carrying a dummy LOK launched on 27 June 1971 but failed shortly after lift-off. An amped-up N1 with high-energy cryogenic upper stages was suggested as a means of delivering a lunar lander descent stage into lunar orbit. This would be followed by a three-man lander/Earth return vehicle launched by another N1. Discussion of lunar stays of up to four weeks was envisioned but no funding was forthcoming. A fourth N1 that had been extensively redesigned launched on 23 November 1972 but 90 seconds into flight, the rocket was destroyed by range safety following a fire in the first stage. The Luna program was scaled down, although more N1s were built for a tentative attempt to land a cosmonaut on the Moon in the mid-1970s. But, in 1976, the N1 program was finally killed and work began on developing the Soviet Shuttle. The N1? The remaining six launch vehicles were destroyed and some components were repurposed for use in the Energia program. Some of the NK-33 (Figure 2.4) engines that provided the power to

34 Government Anchors

2.4 The Soviet Union's NK-33 engines. Credit: NASA

launch the N1 were sold to American companies in 1996 while the LK landers and the LOK found their way into museums and space institutes.

Why did the Soviet lunar program fail? Perhaps it was the lack of funding. After all, Apollo was funded to the tune of US$24 billion whereas the Soviet program had to make do with just US$4.5 billion. Or perhaps it was the lack of a coordinated and cooperative effort between the design bureaus (26 of them!). Perhaps the Soviets entered the race to the

Moon too late? Or perhaps it was just a series of technical shortcomings such as the failure to ensure reliable thrust stability across all the NK-33 engines. In all likelihood, the reason the Soviets failed was due to the combination of the above problems. So how might a Russian Manned Moon Mission 2.0 pan out?

In August 2014, Igor Mitrofanov of the Russian Academy of Sciences Space Research Institute was quoted at the COSPAR Scientific Assembly as saying a manned Moon mission would cost about 100 billion rubles, or about US$2.8 billion. Based on Mitrofanov's announcement the Russian manned Moon program will be completed in five or six years and will culminate with the establishment of lunar bases within 10 years. Kick-starting the return to the Moon will be the launch of Luna-25 in 2016. Luna-25 will land at the Moon's South Pole while follow-on missions will include a lander to search for water ice and an orbiter. While this sounds encouraging, enthusiasm must be tempered by the timeline, since the orbiter is slated for a 2018 launch and the lander for lift-off in 2019. A sample-return mission will be deployed to the lunar surface in the 2021–22 time frame. And the manned landing – that won't take place until 2029–30, by which time there is a good chance Elon Musk will be living on Mars. In preparation for this new armada of lunar vehicles, Roscosmos (see sidebar) is preparing the way by laying the groundwork for constructing a super-heavy lift rocket and advanced manned transportation system (Figure 2.5).

> *Roscosmos, ESA, and NASA*
>
> In January 2012, Roscosmos was in discussions with the European Space Agency (ESA) and NASA about establishing a manned research base on the Moon. Why the offer of collaboration? Perhaps it was funding. After all, any lunar enterprise will come with a hefty price tag. Or perhaps the Russians were losing confidence in their technical prowess following a run of bad luck that included a malfunctioning Phobos-Grunt Mars vehicle and a series of botched satellite launches?

Will Russia be successful in establishing a base on the Moon? Well, it will depend on Roscosmos having a heavy lift launch vehicle – something the Russians haven't had since the cancellation of the Energia booster. Today, as Russia watches NASA and the Chinese build their own heavy lift launchers, Roscosmos is firmly committed to building its own super booster. The next step will be developing a vehicle (Figure 2.6) equivalent to NASA's Orion and, assuming they can deliver on this, all that remains is developing a lunar lander. Russia still smarts at having been beaten in the first Space Race to the Moon and it certainly won't want to end up being Number 2 or Number 3 (behind China) in Space Race 2.0.

LUNAR NATIONS: CHINA

China is moving aggressively ahead in its manned exploration program and one of its goals is a manned lunar mission. From robotic missions to single-piloted missions to multiple crews to spacewalks to space stations, the Chinese have taken a steady approach to

2.5 A model of Russia's planned heavy lift launch vehicle. Credit: FlightGlobal, David Todd

realizing an optimistic but reasonable plan. Having taken the best of the technology they imported from the Russians and having learned from what ESA and NASA have achieved, the Chinese are slowly but surely building a robust manned space program and, while getting to the Moon won't be easy, the Chinese are very serious about achieving that goal. And the signs are everywhere. In 2020 or thereabouts, China will orbit its own space station – a

Lunar nations: China 37

2.6 The TMA Soyuz. A beefed-up version of this will be needed to transport crews to the Moon. Credit: NASA

move that will not only give the nation a definitive edge in space, but also lay the foundation for an eventual manned lunar mission. And when Chinese taikonauts (Figure 2.7) embark on their lunar voyage, it won't be for a "flags and footprints" mission because China's program is deeply rooted in realizing a global strategic and economic footprint: China will be going to the Moon to mine lunar reserves such as titanium and helium-3, which just happens to be the perfect fuel for fusion reactors.

China's march into space: a snapshot

1970: Launch of Dongfanghong 1, China's first satellite; it orbits for 26 days
1975: Launch of China's first remote-sensing satellite
1999: Shenzhou 1, the first of several unmanned spacecraft, is launched
2003: Shenzhou 5 launches, carrying China's first taikonaut, Yang Liwei
2007: Chang'e 1 is launched, the first of a series of lunar probes
2008: Zhai Zhigang performs China's first spacewalk when he steps out of the Shenzhou 7
2011: Unmanned vehicles Shenzhou 8 and Tiangong 1 dock
2012: China launches its first female taikonaut, Liu Yang, and performs its first manned docking

2.7 Liu Yang in 2012. Yang was selected as a crewmember for the Shenzhou 9 mission that was launched on 16 June 2012. Credit: CNSA

Phase 1: orbiting the Moon

China's lunar program started with the Chang'e 1 lunar orbiter which was launched on 24 October 2007. Chang'e 1's job was to generate high-definition 3D images of the lunar surface to identify potential landing sites for subsequent missions and also to map the distribution of chemicals in the lunar regolith to determine how valuable such resources might be. Chang'e 1 was followed by the launch of a second orbiter, Chang'e 2, which was launched atop a Long March 3C rocket on 1 October 2010. The second of China's lunar probes was tasked with mapping the Moon in greater detail than Chang'e 1 and to test communications from the vicinity of the Moon and deep space.

Phase 2: landing on the surface

Then, on 1 December 2013, China launched Chang'e 3, which carried a lander and a rover. The vehicle landed on the Moon less than two weeks later, where it deployed a 140-kilogram rover named *Yut*u which went about its business of exploring a three-square-kilometer area over the next three months. Next in line was Chang'e 4 but, before that mission, China sent a test capsule around the Moon. Unofficially dubbed Chang'e 5 T1, the vehicle was a pathfinder mission designed to assess technologies for the Chang'e 5 mission planned for 2017. Launched by a Long March 3-C rocket on 23 October 2014, the capsule, named *Xiaofei*, executed its sling-shot trajectory around the Moon before returning to Earth on 1 November at 40,000 kilometers per hour. It marked the first return trip to the Moon in nearly 40 years and China became just the third nation to perform the feat. Next in the sequence of lunar probes is Chang'e 4, which is expected to land a second lunar rover and lander in addition to testing techniques required for the sample-return mission planned for 2017.

Phase 3: sample return

Chang'e 5, which looks suspiciously like a scaled-down version of a manned descent vehicle, will be launched by a Long March 5 rocket sometime in 2017. In addition to duplicating the re-entry of the *Xiaofei*, Chang'e 5 (Figure 2.8) will also be tasked with collecting two kilograms of lunar rocks, launching from the lunar surface and rendezvousing with a return module in lunar orbit. If all goes well, Chang'e 6 will follow Chang'e 5 sometime in 2020. If both missions succeed, it will be yet another important step towards realizing the dream of putting boots on the Moon sometime in the 2025 time frame.

Phase 4: boots on the Moon

Can China do it and, if they can, will they beat the Americans? Well, if we're talking about NASA, the answer is a definite "yes." With continuing budget cuts and a lack of

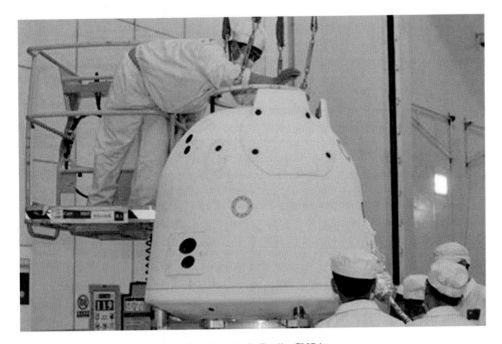

2.8 Chang'e 5. Credit: CNSA

exploration direction beyond the 2023 Orion lunar mission, the American agency will most likely be playing a catch-up game in the manned Moon stakes. And even if a change in government policy targeted a manned lunar landing, chances are the next administration would kill it (as happened with the killing of Constellation by President Obama). The advantage the Chinese (see sidebar) have is that they are unencumbered by new administrations sabotaging bold exploration endeavors – China makes a plan and then executes it, as evidenced by their remarkable track record of spaceflights to date. While NASA dilly dallies between landing astronauts on asteroids and trying to appease the "Humans on Mars" crowd, China slowly and surely moves towards *cis*-lunar dominance following a systematic and precisely executed path towards realizing a permanent base on the Moon. Now, if we're talking about Bigelow (see Chapter 3), then the money has to be on the Budget Suites of America owner to be first back to the Moon, but more on that later.

> *Manned mission to the Moon?*
>
> In April 2015, China's chief designer of its space program, Zhou Jianping, stated that, although his country had the technology to ultimately achieve a manned lunar mission, China had no plans to land its taikonauts on the Moon. Perhaps, but its recent spate of lunar probes tell a different story.

LUNAR NATIONS: EUROPEAN SPACE AGENCY (ESA)

Okay, so ESA is not a nation but a consortium of countries that represent a government-funded effort to return humans to the Moon. Perhaps. ESA began studying missions to the Moon in 1980 when the Polar Orbiting Lunar Observatory (POLO) was on the table. POLO was to have comprised an orbiter and a relay satellite and would have been deployed from a Shuttle or launched by an Ariane. POLO was never flown but, 10 years later, the Moon Orbiting Observatory (MORO) mission was developed. MORO was a medium-sized scientific mission that would have been launched on an Ariane-5 into geostationary transfer orbit (GEO). From there, a translunar orbit would have been used to travel to the Moon and the vehicle would have been placed into a circular polar orbit. Unfortunately, MORO didn't fly either. Then, in 1994, at the first International Lunar Workshop in Beatenberg, Switzerland, ESA submitted a proposal for a four-phase Moon exploration program. Phase 1 would comprise robotic explorers, Phase 2 would ensure a permanent robotic presence, Phase 3 would utilize lunar resources, and Phase 4 would wrap up the program by landing astronauts on the surface. The workshop generated some interesting ideas about the best use of translunar injection and translunar orbit injection, but achieved little in terms of progressing to the stage of implementation. Another assessment study was conducted by the agency of the Lunar European Demonstration Approach (LEDA) which defined a mission to land on the lunar surface following an Ariane-5 launch. Following LEDA was EuroMoon 2000, which was a mission planned for the lunar South Pole. The 1,300-kilogram EuroMoon spacecraft would have been launched by the end of the millennium on a Soyuz but the venture was cancelled in March 1998. A more successful mission was the Small Missions for Advanced Research in Technology-1 (SMART-1), which was launched on 27 September 2003. Just one meter across and weighing 367 kilograms (of which 287 kilograms was non-propellant), SMART-1 (Figure 2.9) used a Hall-effect thruster to travel to the Moon and cost just US$170 million. Despite its small size, SMART-1 was packed to the gills with experiments, including cameras for lunar imaging, an X-ray telescope, an X-ray solar monitor, an infrared spectrometer, and a dust experiment.

After its launch from French Guyana, SMART-1 was inserted into a geosynchronous transfer orbit (GTO) of 7,035 by 42,223 kilometers, and its Solar Electric Primary Propulsion (SEPP) kicked in and began the task of accelerating the vehicle towards the Moon. Since the electric thruster didn't provide as much power as conventional chemical propulsion, it wasn't until February 2005 that SMART-1 finally decelerated into its final orbit above the Moon. Once there, it spent its mission conducting its mapping and chemical analysis tasks. Its mission came to a spectacular end on 3 September 2006 when the spacecraft de-orbited and hurtled towards the lunar surface at 2,000 meters per second. One of the reasons for ending the mission destructively was to simulate a meteor impact and also to expose materials in the regolith such as water ice. When SMART-1 crashed into the Moon, the impact was so violent that the event was observable by ground-based telescopes such as the one on Mauna Kea in Hawai'i. It was estimated the vehicle left behind a three-by-ten-meter crater on the lunar surface and that debris was spread over 80 square kilometers. SMART-1's legacy? Well, it was a super efficient spacecraft that traveled 100 million kilometers using just 60 liters of fuel. The vehicle also generated detailed

2.9 The European Space Agency (ESA)'s SMART-1. Credit: ESA

mineral maps of the Moon, identifying distribution of iron, magnesium, calcium, and aluminum.

After the destruction of SMART-1, there wasn't much talk about returning to the Moon until the beginning of 2015 when the subject of a manned mission was announced by the media. The first clue that ESA had intentions of landing its astronauts on the Moon was the release of the *Destination Moon* video in January 2015 (www.esa.int/spaceinvideos/ Videos/2015/01/Destination_Moon). In the eight-minute video, ESA describes broadly how the agency may visit the Moon in preparation for missions further afield. While no references to specific missions were made, the video mentioned the Lunar Lander (Figure 2.10) mission that had been put on hold since 2012 due to lack of funding. Now, with funding available, it appears as if the Lunar Lander will go ahead in 2018, when it will visit the lunar South Pole to search for water.

In addition to searching for water, ESA's Lunar Lander will be tasked with testing the technologies required to land with precision and to autonomously avoid surface hazards such as boulders and craters. During its surface stay, the Lander will also characterize the lunar radiation environment and also the effects of the fine lunar dust that caused so many problems during the Apollo missions. This latter issue will be particularly important because, as the Apollo astronauts discovered, lunar dust gets everywhere, covering solar panels and causing all sorts of mechanical problems. The Lander will also investigate the electrostatic properties of the lunar dust because it is possible that electrostatic

2.10 The European Space Agency (ESA)'s Lunar Lander. Credit: ESA

charging could prove hazardous to astronauts. And, in addition to all those tasks, the Lander will search for resources such as ice, metals, and nitrogen that could be used for future manned missions.

Here's how the mission may play out. Sometime in 2018, the 2,500-kilogram Lunar Lander will be launched on a Soyuz rocket from Europe's spaceport in French Guyana. After being injected into an elliptical orbit by the launcher's Fregat upper stage, the Lander will use its own propulsion system for the rest of its journey to the Moon. Once in polar orbit around the Moon, the Lander will maneuver itself into a circular orbit 100 kilometers above the lunar surface. Once ground tracking has determined the vehicle's velocity and position and once all systems are confirmed to functioning nominally, the Lander will begin its descent to the surface. First the Lander will fire its engines over the North Pole and coast for half an orbit, during which time the vehicle will match landmarks against those stored in its memory to determine its position. As it approaches its destination, the Lander will fine-tune its thrust while simultaneously scanning the surface for hazards such as boulders and craters. Since there will be no time for ground intervention the final descent will require full autonomy. A few seconds prior to touchdown, the Lander will deploy its four legs and come to rest in the South Pole region at the edge of the Aitken Basin, a possible location for a future manned mission. Once on the surface, the Lander will deploy its high-gain antenna and establish communication with Earth directly, as there will be no relay orbiter in this mission. During its stay on the surface the Lander will rely on solar power to operate its suite of instruments that will be used to search for water and volatiles.

What will happen after the Lunar Lander? Well, it depends on funding and political viscosity, but ESA has some interesting technologies being developed that will be required if and when the agency sends its astronauts to the lunar surface. Perhaps the most interesting of these technologies is 3D printing – a technology that ESA hopes will be the solution to creating a lunar base:

> "3D printing offers a potential means of facilitating lunar settlement with reduced logistics from Earth. The new possibilities this work opens up can then be considered by international space agencies as part of the current development of a common exploration strategy."
>
> *Scott Hovland, ESA Human Spaceflight Team*

SinterHab

Project SinterHab is a proposal put forward by Architecture Et Cetera (A-ETC), a team of space architects that comprises Tomas Rousek, Katrina Eriksson, and Ondrej Doule. The project was kick-started by the three space architects in collaboration with Richard Rieber at NASA in 2009, where Rousek completed an internship. While at NASA, Rousek also had the opportunity to work with Scott Howe, one of the developers of the 3D robotic printing system. Very simply, A-ETC envisions constructing a lunar base from lunar dust using microwaves and solar energy. The space architects presented their idea in March 2013 – a vision (Figure 2.11) that revealed the potential of 3D printing technology to manufacture modules from lunar soil using the sintering process.

2.11 The European Space Agency (ESA)'s 3D printed SinterHab. Credit: ESA

Science fiction? Not at all. Microwave sintering has been around for a while and experience has demonstrated that the process can be applied to sinter all sorts of materials in powdered form: everything from cobalt to carbide and from tungsten to tin. So why not lunar bases? Here's how it may work. First, the nano-sized particles of lunar dust would be heated up to 1,200°C before being melted in a microwave oven. During the melting phase, the regolith particles would bond together, creating building blocks which would be used to build the lunar base. No need for glue or binding agents and no need for astronauts because this sintering process can be entirely automated, which is great news because it mitigates the health hazards of lunar dust. The automation in question happens to be a six-legged robot dubbed the All-Terrain Hex-Limbed Extra-Terrestrial Explorer (ATHLETE) (Figure 2.12) and it would be the ATHLETE's task to operate the Microwave Sinterator Freeform Additive Construction System (MS-FACS), which is NASA's mobile 3D printer system.

First, the regolith would be excavated by a bulldozer version of the Chariot rover (Figure 2.13). The lunar dust would then be processed by the MS-FACS before being fed to the ATHLETE, which would complete the construction. Sounds like a great plan in theory, but can it work in practice? Well, scientists at Washington State University have tested the method and the results are promising. Since building a base constructed entirely of building blocks would be a lengthy process, the base will probably be a hybrid structure comprising the sintered blocks and inflatable membrane structures (Figure 2.14).

The architectural concept that you can see depicted in Figure 2.14 is one created by renowned architects Foster and Partners, and the location of this hypothetical base would be the rim of the Shackleton Crater at the South Pole. Could it happen? Possibly. While the sintered lunar base remains a drawing-board concept, the additive manufacturing process

2.12 NASA's multipurpose ATHLETE rover. Credit: NASA

is gaining more and more interest, as evidenced by the 350 scientists and engineers who gathered in Noordwijk in November 2014 to discuss the potential of 3D manufacturing. The interest bodes well for not only a manned mission to the Moon, but also an eventual expedition to Mars because the lunar surface will be the ideal proving ground for establishing an outpost comprising an inflatable habitat.

LUNAR NATIONS: US

"The brief Apollo missions were terminated before we could start investigating the requirements for longer term missions away from Earth. Before we set off to distant destinations, like Mars, it makes sense to use our nearest near-Earth object – the Moon – as a test-bed to see how humans will cope with longer duration periods distant from Earth, including on the surface of another planetary body, and how effectively we can make use of in-situ resources to sustain our presence elsewhere."

Stephen Mackwell, Director of the Lunar and Planetary Institute in Houston

2.13 NASA's Chariot rover practicing lunar excavation activities. Credit: NASA

2.14 Inflatable membrane structures. Credit: ESA

Lunar nations: US 47

In a few years from now, the US will once again have the capability of sending a manned spacecraft to the Moon, and it's all thanks to the Space Launch System (SLS) (Figure 2.15) and Orion. But, although the Moon was the destination during the short-lived era of the Vision for Space Exploration (VSE) and Constellation (see sidebar), the lunar surface doesn't seem to be on the exploration agenda. There is talk about missions

2.15 Space Launch System (SLS). Credit: NASA

to asteroids and eventual missions to Mars, even though the price tag of the latter would probably send most of those in Congress into cardiac arrest. There have been opaque discussions about sending astronauts into deep space but no clear-cut destination has been identified – except the obvious one. Ever since the end of Apollo, strategy documents advocating America's return to the Moon have been published like clockwork. Many of these documents have been generated by the Lunar Exploration Analysis Group (LEAG) (see sidebar) which evolved from the Lunar Exploration Working Group that was formed by NASA following the demise of the Apollo Program.

> "The President made a mistake. Because that is the perception. That he killed the space program."
> Senator Bill Nelson (D), Florida, commenting on President Obama's decision to kill the Constellation program in NASA's 2011 budget

---

Obama lies, NASA dies: the death of Constellation

Constellation, which was announced by President Bush in 2004, was the designated successor to the Shuttle program. Using the Ares 1 and Ares 5 launchers and the Orion spacecraft, American astronauts would have been deployed on exciting missions beyond LEO for the first time in decades. So when the program was cancelled, President Obama was strongly criticized by members of both houses of Congress, prompting some political leaders to accuse the president of abdicating US leadership in space and being the catalyst for the death march of US human spaceflight. At the time of the cancellation, US$9 billion had been spent and another US$2.5 billion was needed to terminate the program. Imagine giving that money to a commercial enterprise such as Bigelow? There would be dozens of bases on the Moon by now. Dozens. But back to Constellation. Of course, as we now know, the program wasn't killed entirely. Senator Bill Nelson fought to stem the hemorrhaging of job losses by persuading the Obama Administration to build the SLS and Orion. The problem has since become what the goals of these vehicles might be.

LEAG states the value of human exploration of the Moon has four primary benefits:

1. As an open gateway to the Solar System, where "lunar resources can be used for fuel and life support for operations in Earth–Moon space and for voyages to Mars and beyond"
2. Enabling new scientific discoveries on the lunar surface to support studies and sample analysis to better determine events within the Solar System and better understand the Moon itself
3. Pioneering development of new technologies in hardware and infrastructure to expand commercial involvement that would also provide benefits on Earth
4. To promote international partnerships to make accesses to destinations beyond low Earth orbit available to other nations

It was the LEAG that was tasked with analyzing the scientific and operational issues with lunar exploration in support of VSE. In their first report, the group deemed a permanent presence on the Moon essential to the human exploration of the rest of the Solar System and stated that a lunar base would serve as an ideal test bed for technologies required for a manned mission to Mars. Over the years, LEAG continued to identify lunar exploration goals and provided rationale for a return of crews to the lunar surface. As we now know, VSE was cancelled after the bean-counters decided costs of attempting a manned return to the Moon was just too expensive given the constrained NASA budget. The result was that NASA put plans for a manned return to the Moon on a back burner and focused their efforts and resources elsewhere. But that didn't divert LEAG's purpose and direction. The group met annually and presented papers at conferences that promoted the return of boots on the regolith. As the years went by, a growing consensus formed that it probably wasn't going to be possible for the US to go it alone when it came to returning astronauts to the Moon. People pointed to the ISS model of cooperation and suggested that, with that same teamwork and commitment, NASA, together with the Canadian Space Agency, the Russian Space Agency, the Japan Aerospace Exploration Agency, and ESA, could pull off a return to the Moon (Figure 2.16). Evidently, NASA took notice because, in 2014, the agency posted a request for proposals from the commercial sector for Lunar Cargo Transportation and Landing by Soft Touchdown (CATALYST). CATALYST, which may well make use of the SLS as a means to return to the Moon, is a topic we'll be returning to in Chapter 3, but for now we'll return to the activities of the LEAG.

In 2013, in an effort to drum up interest among those who make space policy decisions, LEAG petitioned members of Congress, submitting a letter with a flyer with the title "Destination Moon." The flyer stated: "The Moon is the most accessible destination for

2.16 The next government-sponsored manned Moon mission may involve several agencies. Credit: NASA

realizing commercial, exploration, and scientific objectives beyond low Earth orbit." The return to the Moon received another boost shortly after LEAG's petition when Republican Frank Wolf submitted a letter to President Obama calling for the White House to discuss a manned mission to the Moon. Wolf's letter was sent shortly after China's Jade Rabbit (*Yutu*) had landed on the Moon – an event which prompted Wolf to state: "As China prepare to send a series of increasingly advanced rovers to the Moon in preparation for what most observers believe will ultimately be human opportunity, many are asking why the US is not using this opportunity to lead our international partners in an American-led return to the Moon." The letter made little impact with the Obama Administration, which should have surprised nobody given that this was the same administration that tried to kill everything that was Constellation (it was only thanks to the dedicated efforts of certain individuals that Orion survived). Shortly after Wolf's letter, NASA announced it was canvassing interest among those commercial companies interested in returning Americans to the Moon. As with everything NASA does, this initiative had an acronym: the CATALYST. Under this program, NASA said it was interested in collaborating with companies via commercial Space Act Agreement (SAAs) with the goal of creating opportunities to advance technologies needed to explore the Moon. Some of these technologies include resource prospecting, sample return, and lunar resource processing, and, the way NASA sees it, commercial companies could help develop these technologies. While the SAAs provide no funding, NASA will provide technical expertise in addition to state-of-the-art mission simulation and mission design software. This collaborative approach makes sense since it has worked well with SpaceX, but the spectrum of collaboration doesn't include manned systems, although the new generation of robotic landers (Figure 2.17) that may be developed under these SAAs may help accelerate a manned mission. *May*.

I say "may" because, in 2015, the NASA exploration plan, vague as it is, sees astronauts visiting an asteroid. It's all part of the Asteroid Redirect Mission (ARM) in which a robotic probe is deployed to grab an asteroid (Figure 2.18) and bring it into lunar orbit where astronauts can take a peek at the rock. The theory behind this is that such a mission would use the SLS and Orion as well as developing the technologies required for a manned Mars mission sometime in the very distant future.

A tentative timeline places such a mission in the 2025 time frame but this date is likely to slip significantly because no robotic probe has been developed, no mission schedule exists, and no asteroid target has been identified. And, to top it all off, there is significant political resistance to this ARM venture because many members of Congress don't think an asteroid is an inspirational goal worthy of a nation that once landed humans on the Moon. Also, those same members don't see the ARM as a logical stepping stone to a manned Mars mission, which enjoys support across the political divide:

> "I love NASA. I'm devoted to NASA. But I don't think pushing a rock around space is a productive use of their time and scarce resources."
> *Representative John Culberson, 2013, saying what he thinks of the ARM project; Culberson, in common with many Republicans, favors a return to the Moon*

2.17 NASA's Regolith and Environment Science and Oxygen and Lunar Volatiles Extraction (RESOLVE) rover. Credit: NASA

Culberson's remark about scarce resources is especially valid since the cost of the ARM has been estimated at US$2.6 billion. But the Mars alternative faces the same problem: money. SLS and Orion combined will eat up a staggering US$22 billion by the time of the 2023 manned test flight and this figure doesn't consider costs incurred through slipped timelines. But, in addition to the financial challenges, any concerted effort to visit

2.18 NASA's Asteroid Redirect Mission (ARM). Credit: NASA

Mars must also contend with having to re-justify the program to each new administration. It happens all the time: anyone remember VentureStar (Figure 2.19)? Apollo was carried out in just eight years but a Mars mission program will take 20 years or more. That's plenty of time for a new president to come along and decide the program is too expensive. And the experiences of Constellation and VentureStar make the bean-counters extremely wary of committing billions of dollars to a project that may very well get axed in the next administration. That's not to say the Mars plan can't be achieved; it just means it may need reworking.

> "Right now, they say, 'We're going to Mars'. Well, that's great, but they haven't said how we're going to get there. So no one knows what to build, when to build it, or how to pay for it."
>
> *Republican Steven Palazzo arguing for a return to the Moon*

One way to rework the Mars mission and still develop the technology needed to reach the Red Planet is to return to the Moon. That happens to be the vision of Republican Steven Palazzo, the 2015 chair of the House Subcommittee on Space. Palazzo's pitch is to use a return to the Moon as a national security issue, highlighting the fact that Russia and China are pursuing lunar programs. He has a point. After all, while manned spaceflight is popular in most countries, it happens to be very difficult to actually get anything done if you happen to live in a democracy because, despite this interest, people still prefer their tax dollars to be spent on housing and education. That's not a problem China has with its one-party locked-down system though. No challenges of persuading successive administrations to continue an expensive rocket program for the Chinese because everything is neatly laid out in a series of five-year plans. If Constellation had continued, then the

2.19 VentureStar. Now a fading memory, VentureStar/X-33 was a Skunk Works reusable launch vehicle prototype designed to reduce costs of getting payloads to space. Problems arose with the pressurized composite tank that stored liquid hydrogen that ultimately led to the project's cancellation in 2001 after US$1.5 billion had been spent. Two of the four XRS-2200 Linear Aerospike engines were disassembled and one survived. The vehicle is rumored to be in a storage hangar at Edwards Air Force Base. What began as a platform to prove some cutting-edge technology ended up as another expensive white elephant. Credit: NASA

chances are that Americans would have set foot on the Moon again by 2020. But, as it stands in 2015, the chances of a government-led return are many, *many* years down the road. In fact, if we are comparing government programs, the chances are strong that the Chinese will beat NASA back to the Moon, which will push back a Mars mission even further because the overwhelming majority of the space scientists agree that the Moon should be the next destination and here's why. While today's crop of engineers and scientists have plenty of experience working on the Shuttle or the ISS, there are very few who have the experience of working on planetary environments where radiation, gravity, and entry, descent, and landing (EDL) issues are critical. That's why the Constellation program would have been such a boon for a future Mars mission. Not only would this generation of engineers have gained the knowledge needed for future long-duration missions, the program would also have developed the Lander needed to land on the Moon and the Red

2.20 Tranquility Base. Credit: NASA

Planet. Sure, the ARM venture may be cheaper than a return to the Moon but, after looking at a rock, NASA still has to return to Earth and develop and test a lander if it is to go onwards to Mars, so why not just test and develop that lander as part of a Moon mission? It makes sense but when has sense come into the decision process? Will NASA return to the Moon? Not without a seismic shift in policy. So, while the agency (see sidebar) deploys its astronauts to rocks in space, it will most likely be left to the rest of the world to visit the Moon. Bases will be built and resources will be mined. Perhaps the Chinese will wander over to Tranquility Base (Figure 2.20), pull up the American flag, and bring it back to Beijing for display in the China Science and Technology Museum? Hell, perhaps they will bring one of the lunar rovers back with them. Who knows.

> *NASA's lunar exploration plans*
>
> It seems NASA is keeping its options open on a return to the Moon given that it is developing the Regolith and Environment Science and Oxygen and Lunar Volatiles Extraction (RESOLVE) rover which is designed to search for water and water ice. The RESOLVE rover will also be tasked with demonstrating how commercial missions might gather and use lunar resources.

Which nation will be first? That's difficult to say but we know that whichever nation (we're assuming a commercial company will be first) it is will need five core elements: a heavy lift launch vehicle; a manned spacecraft with service module capable of traveling to lunar orbit; a lunar lander ascent vehicle; refined rendezvous, guidance, and docking experience; the political will and money to see the venture through to execution. Based on that assessment you would think the US would be odds on favorite, but cash-strapped NASA is further burdened by political viscosity. ESA perhaps? Possibly, if they partner with the Russians. ESA has never been in the manned spaceflight business so they would need to develop a lot of the technology for a manned lunar mission from scratch. And Russia? Well, the Russian Space Agency isn't exactly awash with funding so, while they have stated they want to execute a manned mission before 2030, that may be a long shot. So that leaves China, which has conducted a seamless series of missions over the past three decades: steady, unrelenting, and progressive missions that may just lead the nation to executing a manned mission to the Moon by the end of the 2020s. With that goal in mind, China is developing the Long March 9, which will be capable of delivering lunar-scale payloads into LEO and translunar injection payloads of 45–50 tonnes. In fact, China checks all the boxes for leading the way to the Moon but the chances are that, by the time taikonauts set foot on the surface, a commercial enterprise will have beaten them to it. As we've seen with SpaceX, commercial enterprise moves fast. Real fast.

> *China: first to return to the Moon?*
>
> China's space program is run by the military and China's interests in a manned mission to the Moon will not just be to mine helium-3; it will also be to exert control over *cis*-lunar space. Not convinced? Take the mission profile of the Chang'e 2 spacecraft. This vehicle traveled into and out of lunar orbit before making its way to a Lagrange point, where it stayed a while before intercepting an asteroid. Textbook space control maneuvering! So all this talk of exploiting resources and conducting science may simply be a smokescreen for the real agenda. The Chinese are also keeping their options open when it comes to developing their manned lunar architecture. For example, while the Soviet Union's Luna sample-return mission flew their samples straight back to Earth, the Chinese will be testing a lunar orbit rendezvous before returning the samples. This isn't necessary unless you happen to be planning a manned mission. Just another example of the vision of the Chinese space program.

REFERENCES

1. Abeelen, L. van den. Soviet Lunar Programme. *Spaceflight,* **36**, 90 (1994).
2. Clark, P.S. Chelomei's Alternative Lunar Program. *Quest*, **1992**, 31–34 (1992).
3. Harvey, B. *The New Russian Space Programme.* John Wiley & Sons, Chichester (ISBN 0-471-96014-4) (1996).
4. Hendrickx, B. Soviet Lunar Dream that Faded. *Spaceflight*, **37**, 135–137 (1995).
5. Landis, R.R. The N-1 and the Soviet Manned Lunar Landing Program. *Quest* **1992**, 21–30 (1992).
6. Yasinsky, A. The N-1 Rocket Programme. *Spaceflight*, **35**, 228–229 (1993).

3

Commercial Anchors

Credit: NASA

"There is no reason it won't work just as well on the Moon. Additionally, in this austere (budget) environment, it only makes sense to leverage private sector investments and capabilities. I think there is a great commercial potential on the Moon."

Mike Gold, Corporate Counsel for Bigelow Aerospace

MOON OPEN FOR BUSINESS

At the end of 2013, Bigelow Aerospace submitted a request for a payload review for a lunar habitat to the Federal Aviation Administration (FAA)'s Office of Commercial Space Transportation (AST). At the end of the following year, Bigelow received a letter from the FAA saying that the company was free to conduct commercial activities on the Moon without interference from any other FAA-licensed company. The letter went on to explain that the FAA would use its authority to ensure that private-sector assets such as Bigelow's habitats would be protected. Bigelow's request for the payload review was submitted because the regulatory environment that governs commercial activities on the Moon is opaque at best and Bigelow wanted to have some administrative clarity before committing millions of dollars to building and landing habitats on the lunar surface.

When the story of the payload was reported in February 2015, some mistook the FAA's letter as government endorsement of property rights on the Moon but this was never the intent. The intent behind the FAA letter to Bigelow was to provide the company with an idea of the regulatory framework that guides commercial activities on the Moon. As the FAA noted, there was still some way to go before the subject of property rights could be discussed because the administration didn't have that authority: it can license launches and re-entries but it has no jurisdiction over low Earth orbit (LEO) or the lunar surface. So the upshot of the FAA's letter was that Bigelow could go ahead and construct one of his inflatable habitats on the Moon and not have to worry about anyone interfering with that habitat. When might that happen? Well, first Bigelow will be sending his Bigelow Expandable Activity Module (BEAM) to the International Space Station (ISS), where it will remain for two years, after which he will be orbiting his BA-330 modules for sovereign clients. Once those are up and running, he will focus his attention on a lunar base which could be in the 2025 time frame.

LUNAR MARKETS

Of course, Bigelow isn't the only one who has a commercial eye on the Moon: Shackleton Energy Company (SEC) want to set up a mining base on the lunar surface, Space Adventures are sending a couple of tourists on a fly-by, and OpenLuna are interested in resource extraction. So, with all this commercial interest in the Moon, how strong is the market? It's difficult to say but commercial spaceflight has been subject to market analysis before, as was the case when the Tauri Group assessed the commercial potential for suborbital flight in a 2012 study ("Suborbital Research Vehicles: A 10-Year Forecast of Market Demand") that was funded by FAA-AST and Space Florida. So let's use their analysis as a template and begin by looking at the potential markets as outlined in Table 3.1.

As the Tauri Group did with their forecast, this analysis is based on three scenarios: the Baseline Scenario assumes there is interest in conducting commercial activity on the Moon and utilizing in-situ resources; the Growth Scenario assumes that resource extraction stimulates greater demand for commercial enterprise on the Moon; and the Constrained

Table 3.1 Lunar markets.

Market	Description
Resource extraction	Extracting lunar resources such as water and helium-3
Tourism	Lunar fly-bys and trips by wealthy adventure enthusiasts
Technology test and demonstration	Using the lunar surface as a test bed for technologies needed to establish a permanent presence on the Moon and on Mars
Defense and security	Identification of near Earth objects (NEOs) and military reconnaissance
Science and exploration	Human adaptation to low-gravity environments, space physics, and biological research
Education	Increasing awareness of and access to space
Support and supplies	Manufacturing, life-support consumables, medical care, transportation
Media	Product promotion and increasing brand awareness

Scenario is based on delays in the development of technology, funding shortfalls, and launch failures. All of the scenarios assume a space policy that supports the development of a lunar commercial enterprise.

RESOURCE EXTRACTION

For about four billion years, the lunar surface has been the impact zone for millions of asteroids that have deposited their contents onto the Moon. In addition to the minerals present in these asteroids, the Moon is known to contain water, which can be broken down into hydrogen and oxygen that can be used as rocket fuel. We also know that the Moon contains helium-3, which can be used as a fusion fuel sometime in the future. And then there is the abundance of metals such as platinum and iron, and rare minerals such as lanthanum and samarium which are in much demand on Earth for use in high-tech products. With all this mineral and metal wealth lying around, it's not surprising that commercial enterprises such as SEC are eyeing a potential windfall by gathering these resources. How will Shackleton and the other companies go about this? Well, they will most likely follow a distinct series of steps. First, they will have to embark on a prospecting mission to find the resources and assign a value to them. Second, they will have to calculate just how much of a particular resource exists and determine the grade of the resource before moving on to the third step, which will be calculating whether it is in fact worthwhile extracting the resource. Once resource viability has been determined to be financially worthwhile, the company will move on to the fourth step, which will involve planning the task of mining itself. Next, the company will need to figure out how to remove all the waste material and actually begin the process of mining and establishing processing plants for recovering the resources. Once this is all done, the company will need to reclaim the land to make it suitable for use at a later date. And, because all this work will be prohibitively expensive to be performed by humans, it is likely to be done by robots (Figure 3.1).

60 Commercial Anchors

3.1 Artist's rendition of one of Shackleton Energy Company (SEC)'s robots going about its business. Credit: Shackleton Energy Company

In fact, humans only really come into the picture when full-scale mining operations begin so the first commercial trips to the Moon will likely be robotic missions. Where will they land? Well, one of the most likely destinations will be the South Pole near the Shackleton Crater – an area that receives almost continuous sunlight and a location that contains plenty of water which can be used for fuel and life-support consumables.

TOURISM

The first tourist flight to the Moon will probably be the fly-by mission planned by Space Adventures, which is selling two tickets for US$150 million apiece. At the beginning of 2015, the word was that both tickets had been sold and it is now just a case of upgrading the Soyuz. One of the first items to be upgraded will be the heat shield because the re-entry speed from the Moon will be much higher than the re-entry speed from the ISS. The Soyuz will also need a new habitation module otherwise it will awfully cramped for the return trip. If all goes to plan, sometime in the next few years, two tourists and a pilot will fly to the ISS on a Soyuz. They will stay at the ISS for a few days to do some sightseeing before undocking and rendezvousing with the habitation module and a propulsion stage that will send the Soyuz vehicle on its trajectory to the Moon and back to Earth. Total mission duration is expected to be around 17 days and a tentative launch date is 2017, maybe 2018. While the Space Adventures lunar fly-by will certainly put lunar tourism on the map, it is likely to be a one-off mission, so what other options might there be for the extraordinarily wealthy to visit the Moon? Lunar hotels perhaps? Bigelow is in the business of building space habitats and, while he has distanced himself from the subject of space tourism and hotels in space, there is nothing preventing any company from buying a Bigelow inflatable and fitting it out as a lunar hotel.

Lunar hotels could happen but it will take a while because the very earliest we will see boots on the Moon again will probably be the mid-2020s and these boots will belong to lunar prospectors. And lunar tourism really can't start until there are the necessary facilities in place and, for that, a mode of transportation is required. To begin with, there will need to be passenger services to LEO. Once that is established, say by the late 2020s, lunar-tourism trips could possibly start by the late 2030s or early 2040s, by which time the cost of flying payloads to the lunar surface should be more affordable.

TECHNOLOGY, TEST, AND DEMONSTRATION

This is really one of the core arguments for returning to the Moon and not jetting off to Mars first. The lunar surface will provide the ideal proving ground to test the myriad technologies required for a manned mission to Mars. By developing the surface transportation systems and surface infrastructure on the lunar surface, those embarking on a manned Mars mission will do so knowing they will be applying the highest margins of safety thanks to the lessons learned during their stay on the Moon. While the physics of the return to the Moon are the same as the Apollo mission, this time the goal will be to establish a permanent base, and to do that it will be necessary to test technologies that permit astronauts to live off the land. By doing this, the new generation of lunar explorers will gather invaluable experience for those embarking on a trip to the Red Planet in the very distant future. And let's not pull any punches here: it will be in the distant future because establishing an outpost on the Moon isn't your Shuttle or ISS program – the technologies needed to provide a safe haven on the Moon must be longer-lasting and more robust than those that have gone before. In short, such a venture requires an evolutionary step in the development of ultra-reliable technology, and testing these technologies (Figure 3.2) on

3.2 A NASA concept rover for clearing regolith. Credit: NASA

the Moon will ensure that designs are refined and improved to the level required to sustain long-term human habitation: life-support systems, robotic rovers, spacesuits, oxygen extraction, guidance sensors, lunar-landing engine throttling, radiation protection – it all needs to be tested and tested again.

And, of course, technology isn't the only thing that must be tested. If we are to go to Mars, humans must be tested as well. As was discussed in Chapter 1, the Moon will serve as the ideal test bed for the human body. Let's review the mission here. First, a crew of four will launch to LEO before visiting the ISS for a few days. Then it will be on to lunar orbit, where the intrepid four will spend six months orbiting the Moon during which time any contingency events will be dealt by this group and this group alone – just as if they were going to Mars. After six months of being bombarded by radiation and spending countless hours on the treadmill, the guinea pigs – because that is exactly what they will be – will de-orbit and land on the Moon, where they will spend 12 months doing exactly what they would do on Mars. After a year on the lunar surface, they will shoot back into lunar orbit and spend another six months simulating their return trip to Earth. Then, almost three years after launch from Earth, the now physiologically scarred group of four would return to their home planet to take up residence as lab rats (Figure 3.3) for the next few years.

3.3 Those who return from long-duration tours on the Moon will be prodded and poked to see how spending time in the hostile environment of deep space affects the body. Credit: NASA

DEFENSE AND SECURITY

"I envision expeditious development of the proposal to establish a lunar outpost to be of critical importance to the U. S. Army of the future. This evaluation is apparently shared by the Chief of Staff in view of his expeditious approval and enthusiastic endorsement of initiation of the study. Therefore, the detail to be covered by the investigation and the subsequent plan should be as complete as is feasible in the time limits allowed and within the funds currently available within the office of the Chief of Ordnance."

Arthur G. Trudeau, Lieutenant General, Chief of Research and Development in a letter (classified Secret) proposing the establishment of a military lunar outpost, 20 March 1959

"Defense and security" can have two meanings. For some it conjures up images of asteroid deflection and for others it means missile bases and weapons ready to be deployed from the lunar surface. Both are correct. More than 50 years ago, the concept of lunar-based bombing was alive and well thanks to the arms race in intercontinental ballistic missiles between the US and the Soviets. Neither country wanted to fall behind in this race and so Project Horizon was born. Project Horizon was an Army initiative that argued it was vital that the US should establish a military base on the Moon to be controlled by a unified forces space command. The proposal, which ran to more than 400 pages, went into great detail about how the outpost would evolve, where it would be located, and how nuclear power plants would be built. All being well, the construction would have begun in 1964 and the base would have been ready to launch missiles just five years later. Plans were made to test nuclear weapons on the lunar surface to send a message to the world about the military power of the US. It may sound like alternative history but, not only was this Project Horizon serious stuff, it was a clear indication of just how far Cold War mania had pushed the US. Here's an excerpt:

"The program to establish a lunar base must not be delayed and the initial base design must meet military requirements. For example, the base should be designed as a permanent installation, it should be underground, it should strive to be completely self-supporting, and it should provide suitable accommodations to support extended tours of duty."

The document, which makes for a compelling read as illustrated by the excerpt above, outlined a plan to send a pair of astronauts to set up an advance base camp. They would be followed by a nuclear reactor (what else?) and a tractor that would be used to move lunar regolith and cargo. Despite the fact no astronauts had set foot on the Moon, the scientists and engineers who proposed Project Horizon reckoned it was just a matter of time before the technology existed. Here's another excerpt:

"Ultimately, plant wastes and algae can be used to feed poultry, which thrive in confinement and are, relatively efficient energy converters, producing fresh eggs and meat. Meanwhile, attention will be given to the use of fish and other aquatic animals, such as Daphnia and molluscs, which normally feed on algae."

3.4 An asteroid observatory on the Moon. Credit: ESA

Chickens on the Moon? Yes, the 1960s astronauts would have enjoyed a smörgåsbord of food options unlike the irradiated, sterilized, and processed nutrition that is today's standard fare on board the ISS. Nothing but the very best for the military. Now you may be wondering what happened. Well, the technology just wasn't there (it still isn't) and, after the shock of Sputnik wore off, the US had to divert funds to the Vietnam War. Finally, any possibility of any military presence on the Moon was rendered null and void in 1967 when the US signed the Outer Space Treaty. So that just leaves asteroid defense? Perhaps not, because many space programs are rapidly expanding, and part of that expansion has included greater reconnaissance capabilities, some of which may one day end up on the Moon. But perhaps the most viable commercial role in the lunar-based defense and security arena is planetary defense, which means asteroid monitoring and deflection. There are tens of thousands of near Earth objects (NEOs), some of which have a high probability of impacting our planet in the near future, which means active monitoring will be required (Figure 3.4). And perhaps the best place to do that will be on the lunar surface.

Today, monitoring the risk of NEOs is done at the federal level through programs such as Spaceguard, which is managed by NASA and the US Air Force (USAF). But NASA and USAF can only do so much with the limited funding they have, which means there will be a role to play by commercial ventures as highlighted by Peter Garretson of the US Air Force, who reckons that "the US government will have an encouraging policy to source asteroid-related space-situational awareness from private industry." So, sometime in the near future, commercial companies may be tasked with detection and deflection tasks from the surface of the Moon. Of course, for some commercial companies, the detection of asteroids will serve a dual purpose, since some may be mineral-rich and therefore warrant capture rather than deflection.

SCIENCE AND EXPLORATION

Science can be grouped into five distinct categories: Earth Observation, Space Physics, Astronomy, Materials Sciences, and Life Sciences. Today, most space research is performed on board the ISS, during parabolic flights, or by using sounding rockets. But a lunar location will bring a whole new perspective for space-based research and there are some who can't wait. The International Lunar Observatory Association (ILOA), for instance, is planning a number of missions to the lunar surface, including landing a telescope near the lunar South Pole. Space physics will also be of interest to those commercial entities that establish themselves on the Moon because this is a key part of understanding space weather, which has important implications for humans who happen to be on the lunar surface: a forecast of cosmic rays, solar flares, the radiation environment, and space weather in general will be invaluable to those exploring the surface. Astronomers will be excited by the prospect of conducting their observations from the lunar surface and from lunar orbit because there are many infrared and ultraviolet spectra that cannot be viewed using ground-based telescopes. Likewise, materials scientists will be able to use the lunar surface to conduct research in a long-term microgravity environment – something that is only possible on board the ISS. And the life scientists? They will have a field day – especially if that aforementioned full-length Mars simulation mission goes ahead. But, even if it doesn't, life scientists will be able to study the effect of deep-space radiation on the human guinea pigs (because that is what the first explorers will be) exploring the surface and will be able to test the efficacy of radioprotectants and various other pharmaceutical interventions. They will also be able to assess the effect of the prolonged effects of one-sixth gravity and the health effects of the ubiquitous lunar dust on the pulmonary system.

EDUCATION

Access to the lunar surface, even if it is via a robotic rover, provides unique educational opportunities for universities and schools to increase awareness not only of the Moon, but also of its utility as a stepping stone to Mars. Today, many schools and universities are engaged in the process of developing projects that fly on sounding rockets, parabolic flight campaigns, and even the ISS, so flying projects to the Moon isn't much of a stretch. In fact, this is one of the goals of the Centre for Lunar Science and Exploration (CLSE), which develops undergraduate research programs and channels that into NASA's Science Mission Directorate and Human and Exploration Operations Mission Directorate (HEOMD). These programs in turn provide the field-based training for research on the lunar surface, albeit robotically. But CLSE isn't just about generating research capability, since its activities and the activities of the students it helps create an education and public outreach program that provides material for teachers in the classroom. A good example of this work is the Exploration of the Moon and Asteroids by Secondary Students (ExMASS) program, which is managed by CLSE. ExMASS is a lunar/asteroid research program that features students and teachers working together with scientists. At the end of every year, the students' research is assessed by a group of scientists and the best four teams present their research in a competition to decide who will present their results at the Exploration Science Forum which is held at NASA's Ames Research Centre.

SUPPORT AND SUPPLIES

With all this activity on the Moon, there will obviously be a need for a robust support-and-supply chain. Initially, this network will be used to support robotic missions which will be tasked with establishing the infrastructure necessary for the arrival of astronauts. This means rovers, tools, and in-situ resource utilization (ISRU) supplies must be hauled from Earth to the lunar surface. To get an idea of the extent of the support-and-supply chain during the timeline of establishing a lunar base, it's worthwhile considering the phases of development. To begin with, sites will need to be characterized and mapped. This will be followed by regolith excavation and extraction of volatiles, regolith processing, and the deployment of power sources. Landing sites and communication will also need to be deployed at this stage to prepare for expanding the outpost which will begin with surface construction using more robots and 3D printing. Gradually, a surface transport network will evolve and a basic spaceport will be established. To support the increased human presence, more and more life-support consumables will need to be lugged from Earth, making for a busy supply chain.

MEDIA

Many reading this will have heard of the Mars One boondoggle. As we know now, Mars One isn't going anywhere, but why not apply the same concept to the Moon? After all, if the Mars simulation mission goes ahead, I'm sure there will be plenty of people interested in seeing how astronauts cope with a three-year expedition to the Red Planet, albeit simulated. In addition to the reality TV potential of the Moon, there are more serious applications in the advertising arena, with naming rights for rovers and lunar landers and logos plastered all over the habitats. For example, Bigelow Aerospace offers naming rights for its BA-330 module for US$25 million. Perhaps he'll offer a similar deal for his lunar inflatables? And let's not forget that returning to the Moon will likely be an almost exclusively commercial enterprise so why not have astronauts walking around sporting the logos of the companies that employ them? Just a thought.

SHACKLETON ENERGY COMPANY (SEC)

"There are billions of tons of water ice on the poles of the Moon. We are going to extract it, turn it into rocket fuel and create fuel stations in Earth's orbit. Just like on Earth you won't get far on a single tank of gas, what we can do in space today is straight-jacketed by how much fuel we can bring along from the Earth's surface. Our fuel stations will change how we do business in space and jump-start a multi-trillion dollar industry."

That's the pitch on SEC's website (www.shackletonenergy.com). And it's a pitch that makes sense because the Moon has a lot of water and, as Shackleton point out, this water could very well revolutionize spaceflight: simply mine the ice, split the oxygen and hydrogen, transform it into rocket fuel, and sell it to whoever is on the Moon. This makes a lot of sense because, even if SpaceX can bring down the cost of launching stuff into orbit to less than US$1,000 per kilogram, that's still a hefty price to pay. But, if a lunar-based water-derived in-space refueling capability exists, then access to the Moon could be transformed. But why Shackleton and not NASA or ESA? Well, governments don't have the mandate to do something like this and, even if they did, it would take a long, *long* time due to the political viscosity that affects all government-sponsored space initiatives. As to the question of the potential market that exists for such a service, Shackleton have done a lot of thinking about this and reckon there are four reasons why such a venture will be successful. First is the supply argument. Thanks to the data returned by the Lunar Reconnaissance Orbiter (LRO) and Lunar Crater Observation and Sensing Satellite (LCROSS) probes, Shackleton know there are billions of tonnes of ice trapped beneath the lunar poles which translates into mining opportunities. The second reason is demand. While no demand exists in 2015, Shackleton reckon that interest in the Moon and manned spaceflight in general is increasing and they want to take advantage of that interest as soon as lowered launch costs enable on-orbit commercial business operations to begin. The business case is the third reason. Mining ice is a green business that should, Shackleton hope (see sidebar), attract investors and provide those investors with a good return, especially as spin-off opportunities evolve. The final reason this might all work is the extraordinarily qualified team that Shackleton have assembled to pull off this mining-on-the-Moon enterprise (Figure 3.5).

3.5 Artist's rendition of Shackleton Energy Corporation (SEC)'s mining-on-the-Moon enterprise.
Credit: SEC

> *Shackleton*
>
> "The history of SEC stretches back to 1995 when data from the Clementine mission was being pored over by Dale Tietz and Bill Stone. Dietz, a lieutenant colonel who had been involved in the Strategic Defense Initiative (SDI), noted that the data indicated there were large deposits of water in the area of the Shackleton Crater. Stone, who had spent time developing designs to repurpose Shuttle External Tanks into propellant depots, noticed something else in the data: the possibility to generate rocket propellant in situ to refuel vehicles operating in and around the Moon. Over the next several years Tietz and Stone put their heads together and devised a plan to make money by delivering propellant to LEO. Their plan gained pace following the success of SpaceShipOne and four years later SEC was founded together with high-tech entrepreneur Jim Keravala. And the name? The name 'Shackleton' was used because the great man was the very definition of bold leadership and endeavor and coincidentally, the crater that may be one of the most lucrative mining sites of the Moon bears his name.
>
> "Our analysis shows that low cost fuel in space will be an US$80 billion market by 2030. By then we will have established 20 points of sale in the cis-lunar space, making our presence ubiquitous for all spacefarers in need of quality support services of all kinds. We will be running continuous human-tended industrial-scale operations in space. No one has ever done anything like this before."
>
> *Statement on SEC's website*

So when will all this happen? Well, according to the company's website, Shackleton need to have their infrastructure in place by the mid-2020s to take advantage of the potential lunar market. And the way they plan to do this is in four steps. The first phase is planning, which will involve engaging potential customers and defining the design of the infrastructure (see Appendix I). Once this is done, Shackleton can go about the business of prospecting by deploying robotic probes to identify the most lucrative sites for the base. Next on the agenda will be developing and constructing the infrastructure in preparation for manned operations. With boots on the regolith, Shackleton can then move onto the fourth phase, which is the actual process of mining and selling their product. Of course, this will be very, *very* difficult, because the environment of the Shackleton Crater (Figure 3.6) is very, *very* cold (−200°C), but the great Ernest Shackleton faced tougher challenges and he pulled through. Assuming the first phase is successful and the financial case can be made, the company will begin planning for human mining missions to the Shackleton Crater, beginning with a crew of six to eight who will initially be deployed to LEO for training. Once training is complete, the lunar miners will deploy to the lunar surface and get on with the business of mining operations. And, when the first batch of fuel is processed, it will be sent to a fuel depot somewhere in the vicinity of the ISS, presuming it is still operational (the ISS will be closing its hatches in 2024).

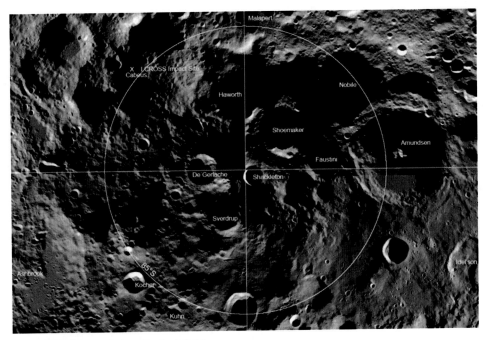

3.6 Shackleton Crater. Credit: NASA

Shackleton's base will most likely make use of the inflatable technology developed by Bigelow over the years. Operationally, crews will rotate out of the base every six months or so, and much of this time will be spent indoors, since robots will take care of most of the day-to-day running of the base. As far as demand is concerned, Shackleton expects this to grow slowly and plan to sell their product to government space agencies and commercial companies for satellite refueling and commercial space operations. One option will be to adopt the Bigelow approach by negotiating agreements with sovereign governments but this may take time given the opaque exploration goals of many of the world's space agencies.

Listen to Jim Keravala and it is impossible not to be persuaded by the business case and the vision but there are many hurdles to overcome before astronauts can turn on the pumps. One challenge will be persuading the US government to allow Shackleton to launch nuclear power supplies (needed for the Shackleton Crater base) to the Moon, and another potential headache will be finding sufficient quantities of plutonium-238 to fuel the nuclear power systems. But perhaps the greatest roadblock is funding, which Shackleton envision to come from private investors primarily. Since Shackleton is looking for up to US$25 billion, they are targeting ultra-wealthy individuals who want to change history and energy consortia who are looking to break into the space propellant business. If US$25 billion sounds like a lot of money, bear in mind that this level of funding is similar to terrestrial large-scale energy projects (Figure 3.7). For example, the development of the Goliat oil field in the North Sea cost close to US$5 billion and the development of just Phase 1 of the Johan Sverdrup field will cost US$18 billion with an anticipated cost of US$33 billion for full field development.

70 Commercial Anchors

3.7 Establishing mining operations on the Moon is an endeavor similar in scale to drilling for oil on Earth. Credit: Divulgação Petrobras/ABr

SHIMIZU CORPORATION

The Energy Paradigm Shift Opens the Door to a Sustainable Society
"A shift from economical use of limited resources to the unlimited use of clean energy is the ultimate dream of all mankind. The LUNA RING, our lunar solar power generation concept, translates this dream into reality through ingenious ideas coupled with advanced space technologies. Virtually inexhaustible, non-polluting solar energy is the ultimate source of green energy that brings prosperity to nature as well as our lives. Shimizu Corporation proposes The LUNA RING for the infinite coexistence of mankind and the Earth."

www.shimz.co.jp

While Shackleton might corner the market for rocket fuel, there are other energy-generating plans being developed, among them a bold concept that envisions wrapping a 20-kilometer-wide band of solar panels around the Moon and beaming power back to Earth. The Shimizu Corporation, which is proposing this audacious plan, reckon their idea could generate 13,000 terawatts of power and provide Earth with an unlimited supply of clean energy. Science fiction? Not quite. The name given to Shimizu's plan is the Luna Ring and, if it is to be realized, there will need to be a number of leaps in the development of robotic technology.

The Luna Ring would stretch around the Moon and would be built by robots. The solar arrays would collect energy from the Sun and this energy would be fed through laser transmitters which would send the energy to receiving stations on Earth. Perhaps it isn't too surprising that the company suggesting this concept is Japanese, since it was Japan that was devastated by the Fukushima meltdown in 2011. Since then, the country has been searching for alternative sources of power and the project concocted by the Shimizu Corporation is about as alternative as it gets. Is it achievable? Not today it isn't, but Shimizu isn't suggesting the Lunar Ring be built today or anytime in the next 10 years. Their plan is to set 2035 as the construction date. By this time, Shimizu reckon, the technology of orbiting solar arrays and microwave transmitters will have evolved to a level such that the Luna Ring is possible. Could it work? Well, let's stretch our minds a little. Let's imagine that space solar power systems (SSPS) are developed in the 2010s and a robotic return to the Moon gathers pace. By the end of the 2010s, robots are scouting around the lunar surface testing ISRU systems and lunar construction methods. By the end of the 2010s, the construction of an international lunar base has begun and, by the early 2020s, pilot demonstrations of SSPS have been conducted from geosynchronous orbit to Earth. By the mid-2020s, the lunar surface is being mined via tele-operation and lunar regolith is being processed for materials production. Then, in the 2030s, commercialization of the Luna Ring begins with placement of a one-gigawatt-class commercial SSPS in geosynchronous orbit. By the mid-2030s, the first steps towards the construction of the Luna Ring begin. Let's wait and see!

BIGELOW AEROSPACE

> "The time has come to get serious about lunar property rights. Ultimately, permanent lunar bases will have to be anchored to permanent commercial facilities. Without property rights there will be no justification for investment and the risk to life."
> *Robert T. Bigelow speaking with NASA manned spaceflight chief William Gerstenmaier*

Robert T. Bigelow wants permission to mine the Moon, his argument being that, without property rights, it will be very difficult to justify the enormous investment and risk needed to set up shop on the lunar surface. But the FAA doesn't have jurisdiction over the Moon and can only regulate launches and rocket re-entries. Another possible legal barrier to mining is the Outer Space Treaty that prevents nations from claiming territory on the lunar surface. I say possible because lawyers have been taking a closer look at the wording of the treaty in recent years and argue that it does allow for property rights. As mentioned earlier in this chapter, a recent ruling by the FAA may help Bigelow with his lunar-mining enterprise. In late 2014, the FAA-AST replied positively to a payload review request submitted by Bigelow. The review was on the subject of commercial development of the Moon and there were some who saw the move as a step towards establishing a legal framework for companies such as Bigelow's to stake their claim on the Moon. This makes perfect sense because, without that legal framework, how are businesses like Shackleton and Bigelow going to attract investors? Any company pondering the investment of billions of dollars will want assurances that their investment will be protected. The FAA-AST decision, which was reached in consultation with NASA and the Department of Defense, provided Bigelow with some of that assurance by saying the AST would do its best to protect commercial assets on the Moon.

It was a step in the right direction, but there is still some paperwork needed to navigate the blind spots in the regulations as they are stated in 2015. One of the first items on the agenda will be to revisit the Commercial Space Launch Act (CSLA) and update it to allow the FAA greater authority for commercial activities such as those that Bigelow is pursuing. The second item will be to have the State Department work with the FAA-AST to define how commercial activities such as mining the Moon will affect current international treaties and this may be a slow process because what Bigelow hopes to do isn't covered specifically. So, while Bigelow can't ask for property rights, he does have the assurance that his activities won't be interfered with. And those lunar-based activities may not be so far over the horizon. As this book is being written, Bigelow is preparing to fly his BEAM on the ISS on SpaceX's CRS-8 flight sometime in the late 2015/early 2016 timeframe. The BEAM (Figure 3.8) will remain there for two years, after which Bigelow will be ready to start flying his sovereign customers to his expandable habitats assuming either or both SpaceX and Boeing have man-rated their Dragon and CST-100 vehicles by then. And, once Bigelow's sovereign customers begin their orbital trips, it may not be long before talk turns to preparing lunar bases.

Based on models presented by Bigelow at various space conferences over the years and the displays presented at Bigelow's shop floor in North Las Vegas, what will likely be the

3.8 Bigelow Aerospace's Bigelow Expandable Activity Module (BEAM). Credit: NASA

very first lunar base will comprise a series of repurposed BA-330 modules. These modules have proven extremely tough and durable, and can easily be modified to feature the flooring necessary for such a base. In addition to the novel use of expandable module technology, the lunar base Bigelow envisages has the capability to descend to the surface thanks to a propulsion bus being attached to the axis, although the specifics of this mode of landing have yet to be field-tested.

OPENLUNA FOUNDATION INC.

"Your Moon, Your Mission, Get Involved!"

www.openluna.org

Another potential commercial route to the Moon is via OpenLuna (see sidebar), a foundation that is targeting the establishing of an eight-man base as its eventual goal. It aims to achieve this through private enterprise in a series of steps, starting with robotic missions and finishing with a manned outpost to be used by government agencies or companies.

OpenLuna

OpenLuna is led by Project Director Paul Graham, an engineer who has Mars analog field experience working for the Mars Society's mission support during the 2002 FMARS season. In addition to developing analog habitation and designing analog missions, Graham also has extensive experience in the design and manufacturing of pressurized habitats. Working alongside Graham is the foundation's chief engineer, Dr. Gary Snyder, who spent several years assessing the strengths and weaknesses of the Mars Direct mission architecture. In addition to Graham and Snyder, OpenLuna's board also includes the Vice President of Membership and Development Debi-Lee Wilkinson, and Kelly Sands, who designs the graphics and illustrations supporting the marketing of the mission plan.

What makes OpenLuna unique in the arena of commercial lunar enterprises is its open-source approach which involves students, education institutions, and the media. It's a strategy designed not only to reach the maximum number of people, but also to provide inspiration to the next generation. With an eye towards saving costs, OpenLuna's mission hardware will be as light and as reusable as possible. Their mission architecture can be distilled into five phases, the first of which will be a scout mission comprising several small rovers that will be deposited on the lunar surface by a single lander. Once on the surface, the rovers will go about their business while communicating with a satellite in lunar orbit. Phase Two is designated the "Boomerang" sample-return mission which will be deployed to one of the locations identified by the rovers. In addition to returning up to 200 kilograms of samples, this phase, which will also make use of a rover, will search for water deposits and helium-3, as well as scouting out suitable locations for a human base.

74 Commercial Anchors

Lunar samples will be returned for analysis at the University of Western Ontario before being auctioned off (on eBay perhaps?) as a means of boosting the funding coffers for the next phase. Phase Three is designated the "Pathfinder" mission – a solo manned mission that will land on the Moon to scout out and prepare the base site for the subsequent manned mission. Next, Phase Four will see the Explorer mission land three astronauts together with the materials needed to build a small outpost, and Phase Five will land another five crewmembers who will spend up to two weeks on the surface.

Once on the surface, OpenLuna's crewmembers will go about fulfilling the foundation's scientific objectives, which will include searching for water and determining the in-situ chemical composition of the regolith – information that will be required to make a decision about where to locate a permanent base. Will OpenLuna succeed? Well, it's all down to money and, although the company states it is investigating funding from television rights and technology spin-offs, as of 2015, there seems to have been little progress.

GOLDEN SPIKE

Mount a Lunar Expedition with us .... *It's the 21st Century!*
www.goldenspikecompany.com
Change is on the lunar horizon—led by the private sector.

"The Golden Spike Company is planning to transform human space exploration by putting in place affordably priced lunar orbital and surface expeditions to the only natural satellite of the Earth – the Moon. Golden Spike will further transform human lunar exploration by making these missions participatory expeditions that involve the general public in ways that create exciting new ways to monetize human space exploration."

In 2012, the Golden Spike Company (Figure 3.9) announced a bold initiative, saying they would begin commercial flights to the Moon by 2020 at a cost of US$1.5 billion a

3.9 Golden Spike Company logo. Credit: Golden Spike

ticket. That seems a lot compared to the US$150 million charged by Space Adventures until you realize that the Golden Spike flights include a landing. The mission architecture chosen by Golden Spike is one that sends the Lunar Lander and crew on separate vehicles, in contrast with the Apollo-style flights that launched everything on a single rocket. Here's how a typical mission might play out.

The basic elements of the mission include the Lunar Lander, the crew vehicle, the lunar transport vehicle (LTV), and the service module. The first launch is an Atlas V that carries the LTV into orbit. This is followed by a Falcon 9 that places the unmanned Lunar Lander into LEO. The LTV then joins the Lunar Lander in LEO and the LTV provides the propulsion to send the Lunar Lander on its journey to the Moon, where it enters lunar orbit and waits for the two-man crew to arrive. Why only two crew? The simple answer is cost: flying three astronauts requires a bigger launcher and a bigger launcher is more expensive. And, on the subject of crews, the astronauts flying these missions won't have to be trained to the same standard as professional government-sponsored astronauts because the flight will be fully automated, just like a robotic mission. No need for pilots. That's Phase I. Phase 2 begins with an Atlas V launching an LTV into LEO and another Falcon 9 launching a crew into LEO. The LTV and the crew vehicle link up in LEO and the LTV boosts the crew to lunar orbit where the crew Lander hooks up with the Lander and the crew execute the landing on the Moon. Once they've finished their work on the Moon, the Lander lifts off and joins with the crew vehicle in lunar orbit. The crew then fires the engine to being their journey back to Earth and the crew vehicle splashes down for an Apollo-style landing.

Since Golden Spike's (Figure 3.10) announcement in 2012, the company has been moving steadily forward. They have selected Northrop Grumman to design the Lunar Lander, although whether the company that built the Eagle vehicle that landed on the Moon in 1969 will build the Golden Spike (see sidebar) equivalent may be another matter. Still, it's a big step towards Golden Spike's goal of ferrying an assortment of wealthy individuals, sovereign customers, and research organizations to and from the Moon – up to 25 such trips if all goes well.

Golden Spike

Golden Spike, which was formed in 2010, has a board of directors that includes former Apollo Flight Director Gerry Griffin, NASA's former head of science missions Alan Stern, and former Space Shuttle Program manager Wayne Hale. The spur for creating the company was the cancellation of the Constellation program which had laid out a road map to return astronauts to the Moon. When Constellation was cancelled, Alan Stern brought together a group of industry leaders to determine whether the private sector could pull off crewed missions to the Moon, and Golden Spike was born.

3.10 Golden Spike Lander. Credit: Golden Spike

Although the catalyst for creating Golden Spike was the death of Constellation, the commercial lunar venture will be taking a very different approach to landing their customers on the Moon. Whereas the Constellation program was based on "clean sheet" hardware, Golden Spike will be relying on a more affordable strategy based on using as much existing hardware as possible. Based on the outcome of several feasibility studies, Golden Spike reckon they can accomplish the first lunar landing for around seven to eight billion dollars. Such a low cost can be achieved because the enterprise doesn't have to develop new flight systems, which also means the venture can be fast-tracked. What this means, according to Golden Spike's plan, is to execute three test flights before landing their first crew on the fourth flight. That sounds pretty ambitious for a start-up company, especially since the announcement of the plan to commercialize trips to the Moon didn't inspire the general public as much as had been hoped. But the response from the scientific communities around the world was much more positive, which is why Golden Spike has pressed on with their plan to prepare for a sample-return mission as the first mission in a series of what the company hopes will culminate with boots on the regolith. Will they be successful? It's difficult to say. With a price tag of US$1.5 billion per flight, this venture isn't targeted at the space-tourism market but there are plenty of countries who might want to take advantage of an Apollo-style mission for the sake of national pride. Although the company only started in 2010 and the targeted mission is in 2020, Golden Spike reckon they are on track to achieve their goal but, as with all commercial space enterprises, financing is the potential mission-killer. The company argues that they are developing their business plan in the same way as any company develops a business plan: by first developing the product and then making the sale. And, since the product in this case is a manned mission to the Moon, there is a fair amount that must be done in the engineering and development phases. But Golden Spike are confident that, when the engineering and development are out of the way, they will be successful in selling expeditions before test flights are completed in 2020/21. Since the company is taking advantage of off-the-shelf technology such as the Falcon 9 launcher and the Dragon capsule, the only item of hardware it has to develop is the lander. Fortunately, the US has had plenty of experience building landers (think of all those landers that have landed on Mars), so all Golden Spike has to do is build a crew cabin and attach it to an existing lander system. And let's not forget that everything that Golden Spike is proposing has been done before: this is very doable. An American railroad to the Moon just might be realized.

4

Technology

"If God wanted us to go back to the Moon, He would have given us money."
Anonymous NASA scientist

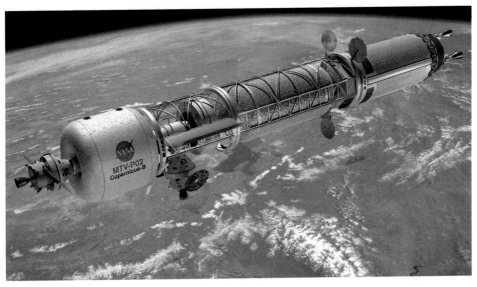

Credit: NASA

When humans finally embark on a mission to Mars, they will need to haul along their life-support system and medical equipment, together with their communications, navigation systems, and myriad other bits and pieces that support an exploration-class expedition. But those selected to venture to the Red Planet won't be able to bring everything that they need, which means they will have to use some of the local resources. This is yet another item that is conveniently overlooked by the "Mars Today" crowd, which isn't surprising because the technology to extract and utilize planetary resources does not yet exist – yet another reason to return to the Moon! But, in addition to testing the technologies needed to live off the land, there are myriad other black holes of knowledge that must be filled before we can even begin to think about deploying humans to Mars. We must develop robust radiation shielding, decide which propulsion system to use, be confident in our hazard-avoidance navigation, know how to deal with the dust problem, and develop ultra-reliable and dependable life-support systems. And that's just a short list. Obviously a discussion of the myriad technologies that need to be developed is beyond the scope of this book, so this chapter provides an insight into some of the big ticket items. We'll begin with transport.

TRANSPORT

In this section, we will discuss the possible mode(s) of transport that may be utilized by those commercial entities setting up shop on the Moon. One option mentioned is the Falcon Heavy (Figure 4.1), which costs US$85 million per launch. This launcher will be capable of delivering 53 tonnes to low Earth orbit (LEO) but only about 10 tonnes to low lunar orbit (LLO), which is slightly less than the throw weight of the Delta IV Heavy (around US$350 to US$400 million per launch) and Atlas V (around US$225 million per launch) launchers. So why go with the Falcon Heavy which has no flight experience, when it is possible to use a launcher with an extensive flight history? Money? Perhaps not. Selecting the Falcon Heavy means tweaking the mission architecture a little, which means a more complex mission design in which the lunar exploration program would have to be assembled in LEO using several launches: one for the crewed vehicle, another for the lander, and another for the translunar injection stage – not as elegant as the traditional approach of launching the whole kit and caboodle on one launcher, and not as safe because increased launches means increased risk. Still, it would be cheaper than the Delta IV Heavy (Figure 4.2) and Atlas V (Figure 4.3) options because neither of these launchers can send the entire lunar exploration program on one launch either. So we'll put the Falcon Heavy on the "maybe" list of options.

Another option for sending the lunar exploration program to its destination is the Space Launch System (SLS). The advantage of the SLS is that it could deliver in one flight what it would take the Falcon Heavy multiple flights to deliver: the Block I will have twice the lunar payload capacity of the Falcon Heavy while the Block II will have up to four times the capacity. But will it be possible to buy four Falcon Heavy rockets for the price of one SLS? That's difficult to say because, as of 2015, the rockets are paper rockets, but we can make a reasonable educated estimation. We know that NASA has stated the estimated development cost of SLS and Orion will be US$30 billion (US$3 billion a year over 10 years) – a figure that does not include operations or payload costs. If the program runs for 30 years, which is the length of time the Shuttle program ran, and there is one SLS launch per year, then the cost of one launch would be US$1 billion. But, added to that figure are the operations costs which may run to US$2 billion a year, which brings us up to US$3

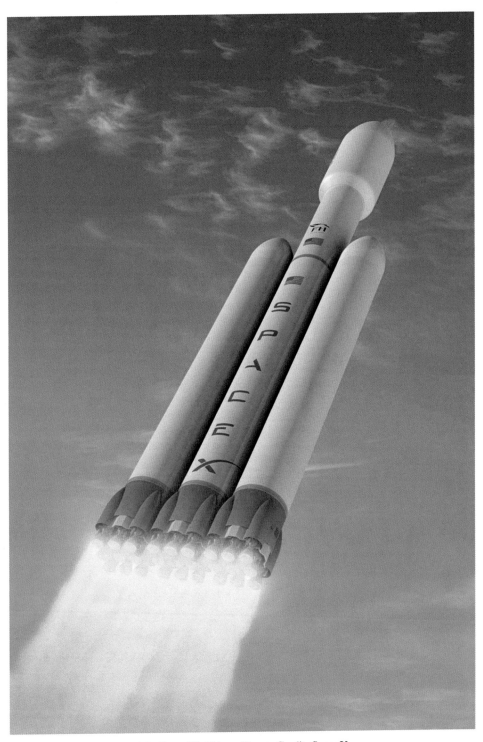

4.1 SpaceX's Falcon Heavy. Credit: SpaceX

4.2 Delta IV Heavy. Credit: NASA

4.3 Atlas V. Credit: NASA

billion per launch. To this number must be added the development costs, which results in a number of more than US$4 billion a launch. And that really strains the NASA budget to such a degree that the agency will be able to develop the rocket but not any payloads for the rocket. In fact, that US$4 billion per launch figure may be optimistic because the launch rate of the SLS may not be one launch every year, but one launch every three or four years. If this is the case, then there are serious concerns about the readiness of the launch team to perform safe launches. As of 2015, the plan is for NASA to develop the Block I version first and complete development of Block II sometime by the end of the 2020s, which is no good if you are a commercial company wanting to set up shop on the lunar surface. If the numbers and NASA documentation are to be believed and the SLS Block I system (Figure 4.4) is ready in 2022 and is used every 4 years, assuming a lifespan of 28 years, the rocket would be launched only seven times, which would equate to US$4.3 billion per launch based on development cost alone. If you add the US$1 billion cost of the SLS first and upper stages to the US$2 billion annual operating costs (multiplied by four remember) and add this number to the one-seventh share of the US$30 billion development costs, you get a figure of US$13.3 billion per launch. That makes the Shuttle appear cheap by comparison. Even if the SLS was launched once per year, it would still be six times more expensive than the Falcon 9 and 15 times more expensive than the Falcon Heavy. As for the cost of launching crews, if the SLS launches once per year, the cost per seat on the Orion is US$88 million and, if it launches once every four years, the cost is US$368 million. Per seat!

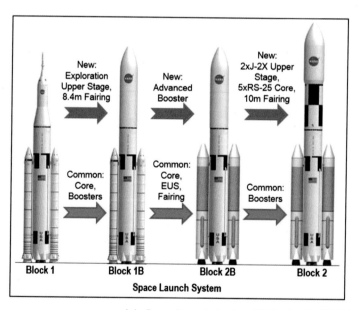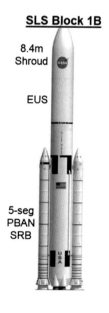

4.4 Space Launch System (SLS). Credit: NASA

Based on these numbers, if you happen to be a commercial enterprise with the goal of starting a business on the Moon, the best bet is the Falcon Heavy route. So that takes care of the launcher, but what about the crew vehicle? The Dragon perhaps? Well, the Dragon has a heatshield that would permit a safe re-entry into Earth's atmosphere, but the Dragon was never designed for a lunar mission, and to upgrade it to lunar status would need some significant tweaking. To begin with, a lander would need to be designed and a service module added. The vehicle would also need an extended life-support system to keep its crew alive for up to two weeks.

HAZARD AVOIDANCE

"Those who forget the past are doomed to land like it. Having looked at the Apollo landings I have come to two conclusions. One – those crews did a great job. Two – data from several of the landings support the idea that we must give future moon landers more information to increase the probability of mission success."

Chirold Epp, NASA's Johnson Space Center

Before their lunar missions, the Apollo astronauts spent a lot of time training to recognize large-scale lunar features near the planned landing site. Not only could these features help the astronauts find their way to the landing site, a familiarity with the topography was necessary because in some cases those features were hazards to be avoided at all costs: hills, boulders, craters – it all helped the crew to find their way before their limited fuel supply ran out. Compounding the navigation challenge was the lunar dust that obscured visibility as the lander made its approach. In some missions, such as the Apollo 12 landing, the dust was such a problem that it completely obliterated the field of view, which meant the pilot – Pete Conrad in this case – couldn't see what was beneath the lander. This was a problem because those Apollo landers were particularly twitchy when it came to the incline of the surface: any steeper than 12° and there was a good chance the lander would be unable to launch back to orbit.

But NASA learned its lessons from Apollo so future lunar explorers should have an easier time finding their way to the surface. Using NASA's Autonomous Landing and Hazard Avoidance Technology (ALHAT), future lunar-bound crews should be able to pilot the lander to a safe and precise touchdown because the ALHAT system is designed to automatically detect hazards. This will be particularly important for missions targeting locations such as the Shackleton Crater because this location is a particularly challenging landing site since the slanting rays of the Sun could hide surface features. To understand how the ALHAT system can do this it is perhaps instructive to imagine a lunar-landing scenario. We'll imagine you're a commercial astronaut employed by the Shackleton Energy Company (SEC) tasked with delivering a small payload to their base. You are piloting the company's lunar landing making your approach from the back side of the Moon, which means it is pitch black. But not to worry because you have ALHAT (Figure 4.5), which uses a suite of sensors that provide you with a continuous stream of data thanks to lasers that can see very well in the dark.

4.5 Planetary landscape designed to test ALHAT. Credit: NASA

By bouncing laser light off the surface as often as 30 times every second, the ALHAT's software processes the information using complex algorithms and generates a 3D view of the lunar surface while simultaneously comparing the lunar features against the surface features stored in its memory. Craters. Mountains. Boulders. It's all presented to you on your screen in crystal-clear definition. It's a technique known as Terrain Relative Navigation (TRN) and, as you slowly close in on the surface, TRN helps you maintain your correct trajectory. Three minutes before touchdown, you are 2,000 meters above the surface but about 100 meters to the right of your intended course. No problem because ALHAT's TRN shows your trajectory angle and descent, and compares them against the ideal trajectory and, because this is an automated system, you can sit back and relax and wait for the ALHAT to kick in and auto-correct. As you check your screen, you hear the soft pulses of the reaction control system correcting and a few moments later the two trajectory lines align. As you pass through 1,000 meters of altitude, the ALHAT's Hazard Detection and Avoidance (HDA) system kicks in and maps all the slopes, boulders, craters, and other hazards near the landing site. After processing the surface features, the HDA provides you with the best landing site option, which appears just as you fly over the Moon's terminator. You are now at 500 meters of altitude and you can clearly see the Shackleton Crater and SEC's base. You keep a close eye on the crater as the altimeter scrolls down the remaining altitude. At this point in the landing sequence, as you check the features visually and compare that with the ALHAT, you enter what human factors engineers term the human interaction interval. You are now at a decision point where you must continue to the intended landing site, abort, or divert to an alternate. But the data stream looks good and you decide to go ahead with the landing at the designated landing site. At 30 meters of altitude, you notice the rocket plumes building until the dust obscures your view of the surface. In the Apollo days, this would have been a problem but you have ALHAT, which continues to generate razor-sharp images. Moments later, the ALHAT navigation screens flashes "Touchdown" – job done.

LIFE-SUPPORT SYSTEM

The first manned missions to the Moon will almost certainly be crews of two or three or four, and the surface stays will be measured in days or weeks, which means the life-support system doesn't need to be closed because most consumables will most likely be

Life-support system 87

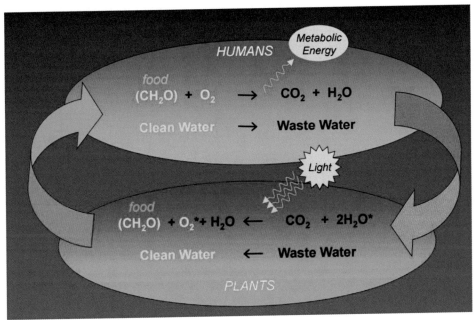

4.6 Cycle of a bioregenerative life-support system (BLSS). Credit: NASA

brought from Earth. But, gradually, as the outpost becomes more permanent, there will be an increasing requirement for a life-support system that is self-sustaining. One such sustainable system is the bioregenerative life-support system (BLSS). A BLSS represents a novel solution for supplementary life support on an extended mission on the lunar surface since it is capable of air revitalization, producing food, reclaiming water, and processing waste [1]. Such biological systems (Figure 4.6) are nothing new, but most of the studies that have envisioned such a system usually imagine large-scale agricultural operations and often optimize just one or two of the many life-support functions without fully taking into account the benefits of using plants. What we're interested in is a small to mid-scale system that can sustain a crew of four to eight astronauts on the Moon.

Before describing how such a system can be developed, it is worth considering the benefits of such a system to the crew. The obvious one is reducing the consumables. Bear in mind that your average crewmember will consume 0.62 kilograms of food, 3.91 kilograms of water, and 0.84 kilograms of oxygen every day. He or she will also produce 5.4 kilograms of waste every day when carbon dioxide, liquids, and solids are considered. Add the consumables per crewmember and you get 5.37 kilograms. Now let's assume there are four astronauts, which brings us to 21.48 kilograms per day, and let's assume they will be away for 1,000 days, which is about the length of your typical manned Mars mission; 1,000 multiplied by 21.48 is 21,480 kilograms, or more than 21 tonnes [2]. That is a horrendous mass penalty! It also represents yet another reason why we must return to the Moon to figure out this BLSS business.

The second advantage of this BLSS concept is nutritional supplementation. Astronauts on board the International Space Station (ISS) must eat packaged food that has been sterilized, thermostabilized, irradiated, freeze-dried, and subjected to just about every type of food processing you can imagine. The result is that it doesn't taste that great and, after a while, eating this pre-packaged food can be a little wearing psychologically. Another downside to eating food that comes out of a packet is that it doesn't have as high a nutritional benefit as fresh food. Let's face it, after spending billions of dollars landing astronauts on the Moon, the last thing we want is for them to suffer malnutrition.

In addition to being able to eat fresh produce, there is great benefit to be had simply by interacting with plants. After all, it is well known that gardening can be a great way to relieve stress and being an astronaut on the surface of the Moon will surely be a nerve-wracking job. We have some experience of the effect of gardening (see sidebar) on the mood of astronauts because some crewmembers have participated in educational experiments to grow food on the ISS. Clayton Anderson, for instance, tended a small lettuce patch and grew basil seeds on the ISS while a group of school students did the same on Earth. Anderson found it a great boost to his psyche when he saw his plants grow – an observation that has also been reported on Earth by employees who work in offices with plenty of plants. And let's not forget that the interior of the ISS is quite cold and technical so plants could add some welcome and soothing color.

> *Plants in space (Silent Running)*
>
> The classic film *Silent Running* takes place at a time when Earth's flora has been destroyed and the only remaining plants are on board greenhouses attached to the spacecraft Valley Forge guarded by off-planet forest rangers, one of whom is Freeman Lowell (played brilliantly by Bruce Dern). When Lowell is ordered to destroy the greenhouses, he rebels and kills his fellow crewmembers in an effort to save the mission. With the assistance of the ship's robots (Huey, Luey, and Dewey), Lowell alters the trajectory of Valley Forge towards deep space, but then receives a transmission that Valley Forge's sister ship has embarked on a mission to intercept the rogue spacecraft. Knowing he will be found out, Lowell jettisons the greenhouse and instructs one of the robots – Huey – to take care of the plants. Then, Lowell detonates the charges and Valley Forge is destroyed. The credits roll with a scene of a greenhouse drifting into deep space tended by Huey.

Another bonus of having plants in an outpost will be the improvement in air quality. Over the years, many astronauts have commented on the less-than-pleasant smell of the ISS – a situation not helped by the high concentration of carbon dioxide (around 2.5% on a good day). But plants can help remove noxious odors and also help remove carbon dioxide from the air in addition to adding oxygen. And then there's the boost from the effect of plants on circadian rhythms: a lunar outpost will be devoid of a normal sunlit cycle and seasons, but plants bloom and wilt and help mark the passage of time.

So what would this lunar-based BLSS look like? Well, it would probably be a tree-based system because woody plants provide food with a low risk of a transfer of pathogens and the roots help break down solid waste very effectively. Also, a tree-based system will provide a variable life cycle with variable growths that will ensure changing scenery and therefore add a psychological boost to the crew. The system would probably be designed to provide 2,800 grams of raw edible plant mass per crewmember per week in accordance with the World Health Organization (WHO)'s recommendations. This amount equates to about 40 servings (one cup) per crewmember per week. Crops will be selected based on the value of micronutrients (Table 4.1) provided by the plants. Not only will these plants have to be rich in folate, calcium, niacin, thiamine, and selenium, they will also have to be nutrient dense, palatable, visually appealing, and easy to integrate into the BLSS.

You can get an idea of the sort of fruits and vegetables that are suited to a lunar BLSS in Table 4.1. The selection of these plants is based on yields, lighting, temperature, carbon dioxide concentration, soil, plant spacing, and myriad other factors. Obviously, in a lunar base, space will be at a premium, so the fruit trees will be of the dwarf variety. For example, a typical dwarf fruit tree will require up to one square meter of space and a soil depth of about 60 centimeters. As mentioned earlier, the plants will absorb carbon dioxide and produce oxygen, which will reduce the burden upon the life-support system's scrubbing system. But plants use a lot of water during photosynthesis and they also lose a lot of water through transpiration, but this water will be collected and condensed before being recycled. Another problem that will need to be resolved is the issue of pests and the effect of reduced gravity on plant growth (Figure 4.7), but work is progressing on these variables.

One potential lunar-based BLSS is described here. Our lunar BLSS comprises three elements: a tree-based waste-management system devised by Tom Watson, known as the Bio-Wick (Figure 4.8); potted plants; and a biodigester. The Bio-Wick, also known as the Watson Wick, processes gray and black water using the symbiosis between woody plants

Table 4.1 Vegetable and fruit growing areas and biomass yields [3].

Vegetables	Growing area (m2)	Edible biomass (g/day)	Inedible biomass (g/day)
Cabbage	0.53	40.00	3.56
Carrot	0.84	62.86	50.29
Celery	0.66	68.57	7.62
Lettuce	0.65	85.71	4.76
Onion	2.23	182.86	50.28
Radish	0.28	25.71	15.43
Strawberry	2.20	171.43	317.98
Tomato	0.79	137.14	100.58
Sweet potato	2.87	148.57	646.34
Fruit trees	*Growing area (m2)*	*Edible biomass (g/day)*	*Inedible biomass (g/day)*
Banana	2	41.95	13.98
Lemons	1	37.73	12.58
Grapes	2	37.20	12.40

4.7 Plant growth on the International Space Station (ISS). Credit: NASA

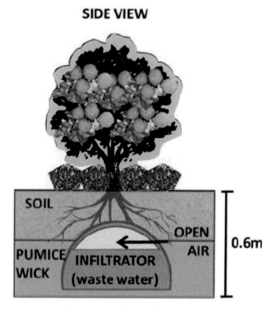

4.8 The Bio-Wick. Credit: NASA

and microbes – a function it achieves thanks to an infiltrator with an open-air volume that is buried beneath a layer of pumice which in turn is buried under soil. This design results in water and waste entering the system and allows air to enter the pumice layer. Thanks to the porosity of the pumice, bacteria can live in the housing and process the waste. Once the bacteria have done their job, nutrients and minerals are drawn up through plant roots and are used to fertilize the vegetation. The Bio-Wick is a novel system that can process all sorts of waste and, thanks to its unique design, there is no need for separate processing systems for black and gray water, as is the case with conventional waste-processing systems.

The vegetable and plants of the BLSS will form the second subsystem that will comprise a series of racks with different plant types occupying separate tiers (Figure 4.9). To reduce the impact on crew time, the maintenance of the racks will most likely be performed robotically, although it is anticipated that the crew will want to water and harvest the plants for reasons stated earlier. The third subsystem – the biodigester – will mix the organic wastes with gray water. The waste material will then be reduced by bacteria, with the resulting product being biogas that could be used for cooking and heating.

4.9 Tiers of plants. Credit: NASA

4.9 (continued)

By now, you may be beginning to appreciate the advantages of a BLSS. First there is the advantage of locating the plants together with the crewmembers, so all the lighting and temperature control is taken care of. Second, plants don't need any replacement parts and, apart from a little water and fertilizer, require little in the way of consumables. And third, there are all those positive psychological benefits mentioned earlier.

HABITAT DESIGN

The first missions to the Moon will almost certainly be performed by crews of four, perhaps six, but regardless of the size of the crew there are some key design factors that will be common to the base that will be established. First, any habitat must be as space- and volume-efficient as possible, which means a dome-shaped structure (Figure 4.10) is probably the most likely design. Here's how such a base might be constructed. First, the base is delivered by a lander, the base module is extracted robotically, and a weight-bearing catenary dome with cellular structured walls is inflated. Lunar regolith is then excavated by robots and this is deposited on the dome to create a protective radiation-resistance shell – a process that takes about three months. With the radiation protection in place, astronauts can move in and begin making the base interior habitable. The dome habitat

Robotics and rovers 93

4.10 An inflatable dome-shaped habitat. Credit: NASA

will be about 10 meters in diameter and divided into three levels: a basement level that will double as a radiation shelter in the event of a solar storm; the first floor which will feature the living area, the plants, and exercise and dining areas; and the second floor which will house storage, the washer and dryer, and the BLSS. Total pressurized volume will be about 440 cubic meters (total habitable space on the ISS is 916 cubic meters), which ensures plenty of space for a crew of four.

An alternative to the dome-shaped base is Bigelow's version of a lunar base. In Bigelow's plan, the entire lunar base would be assembled at one of Earth's Lagrangian points. From there, the outpost, made out of expandable modules, would be piloted by astronauts to the surface of the Moon. No construction on the surface would be required.

ROBOTICS AND ROVERS

With all this talk of digging and depositing regolith onto habitats, it's worth asking how these tasks will be accomplished. The answer is robotics. Whether it be payload removal or stowage operations, or whether geological samples need to be gathered, robots can do it all. But the first task will be to build a base and, for that, a heavy-duty and versatile robot will be required. Which is why NASA designed the All-Terrain Hex-Limbed

Extra-Terrestrial Explorer (Figure 4.11), or ATHLETE (see Chapter 2), a specially designed robot that can deal with just about any terrain.

The ATHLETE is about as versatile a robot as you can imagine thanks in part to those six flexible legs (each of which has seven motorized joints) and wheels. If the terrain is fairly smooth, the rover just rolls along and, if it encounters an obstacle, it can simply step over it. The robot can also divide into two robots that can collect and deliver containers. Thanks to a suite of 48 cameras, ATHLETE can stream video back to its operators on the lunar surface, allowing the crew to check for hazards and to manipulate tools. It can also refill its hydrogen fuel cells, drill for rock samples, and carry a cargo pallet. And, thanks to airless tires, it will never suffer a flat. So the ATHLETE is the robot to use if you need to transport payloads, go searching for geological samples, or to map the local terrain, but what if the crew want to come along? Well, for that, NASA designed a lunar truck called the Chariot. This goes back to the Constellation days but the design is solid and it could easily be resurrected for the next wave of lunar missions.

Chariot, which is as multipurpose a rover as the ATHLETE, comprises a mobility base that features wheel modules, a chassis, and batteries. Like the ATHLETE, Chariot (Figure 4.12) can operate via tele-operation or it can be controlled by a crewmember from the habitation module. With its six wheels, the Chariot is capable of dealing with all sorts of uneven surfaces, and the suspension system ensures a smooth ride for its crew. The rover is also capable of lowering the module to the ground, which allows for ease of entry for the crew and also reduces ground pressure when the Chariot isn't moving.

4.11 The All-Terrain Hex-Limbed Extra-Terrestrial Explorer (ATHLETE). Credit: NASA

4.12 The Chariot rover. Credit: NASA

Chariot's specifications and features include:

- suspension: adaptive suspension: 62 centimeters vertical active travel, 27 centimeters passive travel;
- payload capacity: 1,000 kilograms (crew and cargo);
- steering: all-wheel independent steering; continuous rotation at 145° per second;
- drive: six dual wheels with 20 kilometers per hour top speed;
- capable of climbing a 15° slope in 1 g;
- two wheels per module: neutral and brake;
- crew accommodations: two crew nominal (four for contingency);
- capable of suited crew or shirt-sleeves driving;
- power: 25-kilometer range with 4 hours of run-time.

With its passive and active suspension combined with its ability to lift its wheels independently, this rover is capable of dealing with all sorts of rugged terrain, which it can bulldoze across at speeds of up to 25 kilometers per hour. Since it is capable of generating a pushing and pulling force of more than 1,800 kilograms, the Chariot can be used as a construction vehicle and for various other engineering tasks such as trenching and leveling. In common with so many of the rover's features, the steering system is designed to provide a wide range of steering angles thanks to two gear trains and two motors that permit up to 90° per second steering speed. Power is generated by eight 36 VDC lithium-ion battery packs that ensure a 25-kilometer range over hard ground. And for the crew there is the bolt-on crew short-sleeves accommodation system that can be configured for up to four astronauts or perhaps three astronauts and a Robonaut?

Since its creation at NASA's Johnson Space Center, Robonaut has become something of a celebrity. The dexterous humanoid robot even has its own Facebook site, which had more than 30,000 "Likes" last time I checked. The striking robot has been designed to work alongside astronauts and to do the jobs that are judged too dangerous for humans ... such as constructing a lunar base, for example. The successor to Robonaut – Robonaut 2 (R2) – has already flown in space as a passenger on board STS-133, becoming the first American robot in space (the Canadians have their robotic arm). Highly versatile and extremely dexterous, the R2 (Figure 4.13) features elastic joint technology, force sensing, high-speed joint controllers, and extended finger and thumb travel – features that allow it to compete with humans in the dexterity stakes: one of the tasks it is capable of is replacing an air filter, for example.

For lunar missions, R2 would have its own mobility system for getting around the Moon, which is where the Centaur 2 rover comes in. Developed by the Human Robotics Systems project (part of NASA's Exploration Technology Development and Demonstration Program), the Centaur 2 has been integrated with the R2 (Figure *4.14*), which has resulted in the planet's most dexterous manipulation system. Could R2/Centaur 2 make an appearance on the Moon? Actually, that possibility has already been entertained as a mission known as Project M. This project, which sounds like it could feature in a James Bond movie, was dreamt up after the death of the Constellation program – a decision that

4.13 Robonaut 2 (R2). Credit: NASA

Robotics and rovers 97

4.14 Centaur 2 rover integrated with the Robonaut 2 (R2). Credit: NASA

effectively put an end to government-sponsored manned Moon missions. So some NASA engineers put their heads together and came up with a project that would send a humanoid robot to the Moon in the hope that such a mission might serve as an inspiration for human missions. Once there, the robot would spend 1,000 days (hence the "M" designation, since "M" is the Roman numeral for thousand) traveling around the lunar surface in a rover, deploy surface experiments, demonstrate engineering tasks, and perhaps try its hand at basic construction tasks. It was a great idea and, just as importantly, it was affordable, with an estimated budget of just US$450 million including the cost of the Morpheus Lunar Lander (Figure 4.15) and the cost of launching the rocket, which would be SpaceX's Falcon 9 v1.1 booster.

The Falcon 9 v1.1 can deliver 5,745 kilograms to geosynchronous transfer orbit (GTO) with a delta-v of 2,500 meters per second. Now the total delta-v from LEO to the surface of the Moon is 5,930 meters per second, but this shortfall could be made up by the Morpheus Lander which, based on its specific impulse of 321 seconds, could deliver 350 kilograms to the lunar surface. The Robonaut that would be making the trip wouldn't be

98   Technology

4.15   Morpheus Lunar Lander. Note the ALHAT. Credit: NASA

the R2 but an improved one with legs, dubbed the Valkyrie. And, since the Valkyrie would only weigh 125 kilograms, that leaves plenty of room for surface experiments and sample construction materials. Sadly, this project was never funded, but that's not to say it couldn't be resurrected. What a great way to test and refine engineering and construction techniques and to fine tune the myriad lunar mission management tasks. No life-support systems, no food, no oxygen, and no need for sleep: Valkyrie would be put to work 24/7 and gain invaluable mission-specific experience for the humans that follow. Perhaps Valkyrie could even help the other robots with the construction of the lunar base. How might that be achieved?

STRUCTURES AND CONSTRUCTION

The answer may lie in Contour Crafting, a layered fabrication technology that could build a lunar base in just 24 hours. Contour Crafting is an automated construction technique that uses 3D printing to create large-scale structures from computer-aided design (CAD) models. In the case of a structure such as a lunar base, the walls would be built by extruding a paste-like material similar to concrete and a robotic trowel would be used to create a smooth contoured surface. As you can see in Figure 4.16, the technique is perfect for creating curvilinear shapes and, since the method is fully automated, it reduces not only time, but also cost – ideal for building a lunar base in other words. And this isn't a

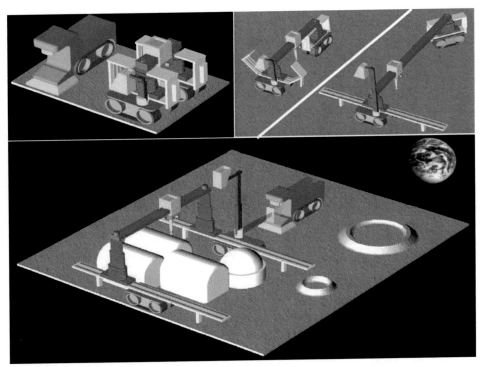

4.16 Contour Crafting.

technology that needs a lot of development before it's ready for prime time. Contour Crafting has demonstrated the feasibility of the technique by building large-scale structures with solid core walls and the technology is versatile enough that it can be used to construct any of the myriad infrastructure elements required for even a basic lunar outpost (Figure 4.16). These elements include structures such as a landing pad, blast walls, a blast apron, a track from the landing pad to the habitat, a platform for the habitat, thermal shade walls, micrometeorite shields, unpressurized shelters for rovers and cargo, and a communication and observation tower.

How will the Contour Crafting system work? Well, it will use in-situ materials, which is the subject of the next section in this chapter, and this is where the challenges are because the structures must have sufficient tensile strength to resist load paths and buckling. The problem is that, until we start building such structures using lunar regolith, we won't know whether it will be possible to ensure the tensile strength of the structures (best to check this out on the Moon rather than Mars though), which means it may be necessary to use Earth-based materials such as carbon fiber as reinforcement. Having said that, the advantage of building on the Moon is that the tensile strength and buckling loads are only subject to one-sixth Earth gravity, which means some of the aforementioned structures can be more slender than their equivalent on Earth, although they will still have to resist meteorites, radiation, and thermal buckling from the temperature extremes caused by diurnal cycling. To cope with this latter issue, it may be necessary to place thermal blankets on the more dynamic structures such as the surface vehicles that stand out in the open. As for the construction material, Contour Crafting reckon they will use regolith binding agents, crushed lunar aggregates, and mechanically shaped rocks as their building materials. Most construction activities will be performed robotically, although some will require tele-operation and minor human supervision.

The first infrastructure elements will be the landing pad, landing apron, and blast walls. Once these structures are built, a road from the landing site to the habitat will be built and this will be followed by the construction of shade walls, a storage hangar, radiation shelter, and the habitat itself. Today's lunar base development philosophy holds that as much local material should be used as possible and this is the construction approach Contour Crafting has adopted. One building material that the company has their eye on is sulfur, which may be used as a binder or perhaps as an ingredient in a regolith-based concrete. Once sulfur is extracted from the regolith, it would be mixed with the regolith at a ratio of 80% regolith to 20% sulfur to produce sulfur concrete, a dry mix that would be extruded through a specialized nozzle barrel heated to 130°C (sulfur's melting temperature). Contour Crafting reckon that the resulting material will have a compressive strength of 3,000 psi, which is stronger than many concrete structures on Earth. So far so good, but the challenge of extruding a granular abrasive material is a significant one made even tougher due to the absence of water, which could act as a lubricant to help the concrete flow [4]. To resolve the problem, Contour Crafting has experimented with a gear motor that turns an augur that in turn forces the mixture into a nozzle. The clogging in this system is prevented by vibrating the barrel when the system senses a clogging event. As to the mixture itself, sulfur concrete, while being a relatively new terrestrial construction material, is very strong and cures very quickly. It is also very resistant to large temperature cycles and it can be molded into straight or curved forms [5, 6]. The only question is how easily local resources can be used.

IN-SITU RESOURCE UTILIZATION

In-situ resource utilization (ISRU) is the extraction and processing of local resources so they can be used to make propellant, life-support consumables, radiation shielding, and, as we have just discussed, habitats. Being able to live off the land (Figure 4.17) is extremely important if we are ever to establish a base on the Moon, never mind Mars, because the cost of hauling stuff to the lunar is obscenely expensive, but that isn't the only reason. Using in-situ resources (Table 4.2) can reduce mission risk by increasing radiation protection, and it can also provide a backup to systems delivered from Earth. As far as a human base is concerned, there are five ISRU areas that must be considered:

1. Resource characterization and mapping
2. Production of mission consumables in situ for power and transport
3. Construction of hardware and crew-protection structures
4. In-situ energy production
5. In-situ manufacturing.

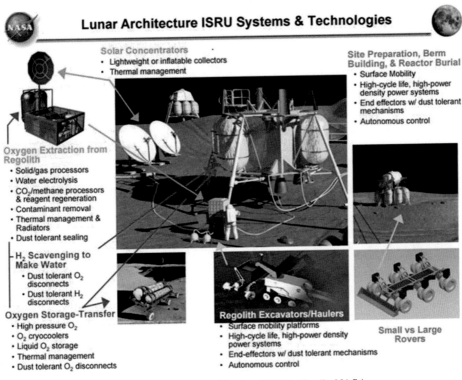

4.17 In-situ resource utilization (ISRU). Credit: NASA

In addition to the basic functions and purpose of ISRU, there are many science requirements that must also be defined. For example, we need to know how to measure the water and hydrogen content in the lunar subsurface, we need some way of measuring the spatial distribution of hydrogen in volatiles that bear hydrogen, and we need some means of extracting subsurface samples. It will also be necessary to measure the geotechnical characteristics of the cold traps on the lunar surface and demonstrate oxygen extraction from the regolith. So there is a lot of work to be done before we can think of sending an ISRU system to the Moon, never mind Mars. Fortunately, work is underway to solve the ISRU challenge and its name is RESOLVE (Regolith & Environment Science and Oxygen & Lunar Volatile Extraction). RESOLVE (Figure 4.18) is a prospecting mission designed to achieve a number of tasks, including confirmation and characterization of the water in materials at the lunar pole and demonstration of the ability to drill without losing volatiles.

At the heart of the mission is the RESOLVE Gen III system depicted atop the Canadian Space Agency (CSA)'s Artemis rover in Figure 4.19. Although the Gen III looks flight-ready, this system was used as part of NASA's third ISRU field test in 2012. The purpose of the Gen III was to develop a flight-ready rover capable of operating on the Moon but, before that could happen, there were a number of tasks that had to be demonstrated

Table 4.2 In-situ resource utilization (ISRU) functions and purpose.

	Tasks and activities	Purpose
1	Resource prospecting/mapping	Map for site selection and ISRU planning
	Chemical and mineral characterization and mapping	Measure and map geotechnical and mineral attributes of regolith
	Map hydrogen and water volatiles near permanently shadowed craters	Measure and map hydrogen and water volatiles to assess potential for extraction
	Characterize solar wind volatiles	Measure solar wind volatiles
2	Consumable production	Reduce Earth delivery logistics
	Oxygen extraction from regolith	Produce oxygen for crew
	Solar wind volatile extraction from regolith	Extract and separate hydrogen and water from regolith
	Water and hydrogen extraction from permanently shadowed craters	Extract and separate hydrogen and water from regolith
	Metal/silicon extraction from regolith	Produce silicon, aluminum from regolith
	Cement for construction	Produce construction material
3	Civil engineering	Reduce mission risk
	Excavate and transport regolith for consumable production	Provide regolith for in-situ processing
	Construct lunar-landing pads	Protect hardware from plume damage
	Utilize regolith for radiation protection	Cover habitat to protect crew from radiation
	Construct structures from in-situ materials	Modify regolith and construct structures for hardware protection
4	Energy production and storage	Enable infrastructure growth
	Construct thermal energy storage from in-situ materials	Modify regolith for use as thermal storage media for energy storage
	Construct solar arrays from in-situ materials	Modify regolith and fabricate solar arrays on lunar surface
5	Manufacturing and reuse	Reduce Earth delivery logistics
	Hardware scavenging and recycling	Remove electrical components from used landers for repurposing
	Prototype part fabrication	Produce spare parts from plastics

terrestrially. The first objective of the field test (held in conjunction with the Moon Mars Analog Mission Activities – MMAMA – mission analog), which was conducted in Hawai'i, was to demonstrate the feasibility of an integrated mobility platform for prospecting for polar ice and volatiles. The second objective was to demonstrate the operation of science instruments and operations when performing investigations of the terrain. In tandem with these objectives, the RESOLVE team wanted to make sure the data, power, and command system integrated seamlessly into the mission profile and also that the basic operations of roving, scanning, drilling, and sample transfer were demonstrated. To that end, the mission was a simulation of an actual lunar mission from landing through to lunar sunset.

During the simulation mission, which was performed at Pu'u Haiwahini, outside Hilo on the Big Island of Hawai'i in July 2012, there were a number of tasks (see sidebar) the RESOLVE team wanted to demonstrate. Controlling initial Artemis rover operations from

RESOLVE Gen III

Purpose: Develop a flight-like unit that can fit on a rover and operate in the lunar environment

Sample Acquisition System
Auger/Core Drill Subsystem [CSA]
- Collect and transfer subsurface material down to 1 m below surface
- Maintain sample stratigraphy and volatiles (below 150 K)
- Meter samples for processing
- Auger material to surface for evaluation
- Measure geotechnical properties of regolith during drilling

Surface Mineral/Volatile Evaluation
Near Infrared Volatile Spectrometer Subsystem (NIRVSS) - ARC
- Measure surface bound OH/H$_2$O while traversing (at min. of 0.5% by mass)
- Detect form of water (ice/hydration) in auger tailings
- Detect water vapor in evolved gases
- Image surface and drill tailings

Resource Localization
Neutron Spectrometer Subsystem (NSS) -ARC
- Locate hydrogen and hydrogen bearing volatiles down to 1 meter below the surface while traversing (at min. of 0.5% by mass)

Volatile Content/Oxygen Extraction
Oxygen & Volatile Extraction Node (OVEN) - JSC
- Accept samples from Sample Acquisition System
- Heat samples from <150 K to 423K for volatile extraction
- Heat samples to 1173 K for oxygen extraction
- Transfer evolved gases to LAVA volatile analyzer

Volatile Content Evaluation
Lunar Advanced Volatile Analysis (LAVA) - KSC
- Accept evolved gas from OVEN; provide hydrogen for oxygen extraction
- Perform analysis in under 2 minutes
- Measure water content in evolved gas
- Characterize volatiles of interest (below 70 amu)
- Measure D/H and O$^{16/18}$ isotopes
- Capture & image water evolved

Operation Control Flight Avionics - KSC
- Space-rated microprocessor
- Control subsystems and manage data

Surface Mobility [CSA]
- Traverse wide range of lunar surface/material conditions
- Tele-operation and autonomous traverse modes
- Carry RESOLVE payload; provide power, comm., and thermal management

4.18 Regolith & Environment Science and Oxygen & Lunar Volatile Extraction (RESOLVE) infographic. Credit: NASA

4.19 The Canadian Space Agency (CSA)'s Artemis rover with the Regolith & Environment Science and Oxygen & Lunar Volatile Extraction (RESOLVE) payload.

a central operations center in Hale Pohaku, control of the rover was later divided between CSA headquarters and Johnson Space Center. Assisting with the mission was the Pacific International Space Center for Exploration Systems (PISCES).

Lunar polar ice/volatile mission objectives

- Travel at least 100 meters on the lunar surface to map the horizontal distribution of volatiles
- Perform at least one coring operation and process the regolith acquired
- Perform one water-droplet demonstration during volatile analysis
- Map the horizontal distribution of volatiles over a point-to-point distance of 500 meters
- Perform coring operations and process regolith at three locations
- Perform three auguring operations
- Perform two total water-droplet demonstrations
- In conjunction with hydrogen reduction, perform one low-temperature volatile analysis
- Perform two coring operations separated by at least 500 meters
- Travel three kilometers in any direction
- Travel directly to local areas of interest
- Process regolith from five cores
- Perform hardware activities that can be used to further develop lunar exploration technologies

---

Why Hawai'i? Well, the Big Island of Hawai'i features terrain similar to the Moon, such as volcanic ash deposits and diverse rock distribution, which not only is ideal for testing ISRU technologies, but also provides a suitable mobility challenge similar to that faced by the rovers that will eventually land on the Moon. The rover in this case was the CSA's Artemis that provided the platform for the RESOLVE suite of instruments that included spectrometers, a drill to take samples, and an oven to bake the samples. In addition to the scientific experiments, the mission also included a group of MMAMA projects designed to test new exploration techniques on the lunar and Martian surfaces. These projects included robotic resource mapping which was conducted by another of the CSA's rovers – Juno – which was fitted with ground-penetrating radar for that purpose. Also tested was a special spectrometer designed to scout out resources in the regolith and below the surface.

The mission simulation was very successful. Seven analog days were completed and most of the mission objectives were demonstrated successfully. The rover rolled off the lander without issue before traversing to its designated locations and executing its tasks (see sidebar). All in all, the RESOLVE demonstrated many of the key ISRU technologies that will be vital to those tasked with establishing a base on the lunar surface and eventually the surface of the Red Planet.

Regolith & Environment Science and Oxygen & Lunar Volatile Extraction (RESOLVE) mission accomplishments

- Tele-operation control of rover in field
- Autonomous traverse of 250 meters (total roving distance: 1,140 meters)
- Site mapping performed with near infrared spectrometer
- Hotspot localization procedures demonstrated
- Augur operations demonstrated
- Sample capture and sample transfer (to oven) demonstrated
- Core segments heated and processed
- Hydrogen reduction demonstrated

NASA was pleased with the results and began laying the groundwork for a Mars-bound rover that will hopefully repeat the objectives of the RESOLVE–Artemis combo on the surface of the Red Planet. In addition to sending a RESOLVE payload, NASA also hopes to add a Resource Prospector Mission (RPM) on a vehicle slated to fly sometime in 2020. But, before the Mars mission, RESOLVE will be headed to the lunar surface in 2018 to scout for areas containing hydrogen and water. If it is successful in its search, it will extract oxygen from the regolith and process it with the hydrogen to make water. This will be big news for lunar and Mars proponents because water means life support and it also means propulsion, which in turn means fuel depots. It will also mark the first time ISRU hardware will have been operated on the lunar surface, but this will only mark the beginning of ISRU in the field. After the proof-of-concept mission, follow-on missions will need to be conducted to demonstrate long-term operations and to assess the impact of the lunar environment on mission hardware. Resource locations will need to be identified, after which integrated missions will need to be performed to demonstrate the feasibility of ISRU power, surface mobility, and power capabilities. ISRU products will need to be demonstrated, after which pathfinder missions will need to be flown to demonstrate the implementation of ISRU into human missions. Many of these missions will be robotic precursor missions and we are still in the early stages of Phase I, which is the feasibility phase. Evolving the myriad ISRU systems will take time and developing those systems to technology readiness level 6 will take even more time. Even if we leave this up to the commercial sector, it may well be the latter part of the 2020s before we see a human ISRU mission on the lunar surface. A dress rehearsal for an eventual Mars mission? Perhaps – if the Variable Specific Impulse Magnetoplasma Rocket (VASIMR) or a bi-modal nuclear thermal rocket is up and running. The late 2020s may sound like a pessimistic forecast, but it's important to consider the scale of what needs to be achieved within the ISRU arena before a human crew can set up an ISRU-fueled base. Systems capable of excavating and transporting material must be developed and proven in the field, end effectors capable of manipulating large and small objects must also be developed and field tested, and oxygen extraction and processing from the regolith must be perfected. Solar concentrators must also be developed, and water, carbon dioxide, and gas-processing systems must also be developed. And all these systems must be dust-insensitive, which means scientists and engineers must develop a thorough understanding of the characteristics of the lunar regolith and the way it behaves when manipulated by excavators and transportation systems. Even with aggressive funding and fast-tracking of development, it is unlikely all these goals can be accomplished before 2030. Now that's pouring cold water on the "Mars Tomorrow" crowd but that's the harsh reality of developing ISRU systems and its best to get these systems right before heading off to Mars.

COMMUNICATIONS

The final technical piece of the manned-Moon-mission puzzle is communications. Once again, by returning to the Moon, it will be possible to test high-precision communications coverage before scaling that technology for a manned mission to Mars. Thanks to Constellation, NASA has already been down this road when they developed a lunar communications architecture for their planned return to the Moon. The architecture for Constellation is still applicable for today's return to the Moon and comprises three phases. The first of these would support robotic missions on the Moon while the second phase would be the return of humans to the lunar surface and the third phase would support a base. The fourth and fifth phases would support a manned mission to Mars. The primary objectives of the lunar communication architecture will be to provide continuous and precise communications and navigational support with minimal interruption.

The first step in developing such an architecture will be to position one satellite in polar orbit and at 90° inclination and another in an equatorial orbit at 0° inclination. These two satellites will ensure preliminary coverage and they will be joined by four more lunar relay satellites to provide almost full communications coverage for the lunar robotic program. Since the satellites would have lifetimes or five years or so, they would obviously have to be replaced at intervals leading up to the human mission and lunar base phases. Part of the communication infrastructure will include the lunar positioning system (LPS) which will enable precise position determination of all the robots and astronauts going about their business. The LPS will have the same function as GPS does on Earth so signals for latitude, longitude, altitude, and time will be required. But four signals don't require four satellites because there are techniques to provide receivers with additional signals. Here's how it might work.

To begin with, three groups of satellites would be placed in orbit around the Moon and Earth. One group would comprise Orbiter satellites that would orbit in two planes, with six satellites in each plane. The second group would be relay satellites, of which there would be two. These would be placed in polar orbit. The third group would comprise three Earth-relay satellites that would be placed in geosynchronous orbit around Earth. This configuration of satellites wouldn't provide complete coverage but the only blind spots would be a short distance from the poles which is of less scientific interest than the actual poles and the equator. With the system up and running, preparations could begin for supporting the second phase – the human mission support phase. This phase would require an enhanced navigation and communication architecture, since astronauts will most likely be moving across the surface more quickly than their robotic counterparts.

For the human mission phase, the communications system would comprise an antenna, a transponder, and a field programmable gate array (FPGA) which would enable fast programmability, beam shifting, and telemetry. FPGAs are reliable and have been a feature of several missions so there is no reason for increasing mission risk by trying a new system. One important feature of the antennae will be beam steering, since this will enable pointing at the relay and Orbiter satellites. Another addition to the second-phase architecture would be adding speech and telemetry channels and boosting the data rate perhaps by means of an optical communication system. The ability to encode data onto a beam of light is a technique NASA has been pursuing for a few years now as part of its Lunar Laser Communications Demonstration program. Using infrared lasers, NASA hopes to dramatically increase two-way communications between Earth and the Moon. The program has already shown signs of success when, in October 2013, the Lunar Laser Communications System (LLCS) used a laser beam to transmit data from the Moon to Earth at a highly impressive 622 megabits per second (Mbps). That's an uplink rate to the Moon that is 5,000 times faster than radio technology. And that's the beauty of optical communication: it can reliably and rapidly transmit huge volumes of data.

REFERENCES

1. Poynter, J.; Bearden, D.E. Biosphere 2: A Closed Bioregenerative Life Support System, an Analog for Long Duration Space Missions. In: Goto et al. (eds), *Plant Production in Closed Ecosystems*, pp. 263–277. The Netherlands, Kluwer Academic Publishers (1997).
2. Barta, D.J.; Henninger, D.L. Regenerative Life Support Systems: Why Do We Need Them? *Advanced Space Research*, **14**(11), 403–410 (1994).
3. World Health Organization. Fruits and Vegetables for Health. Workshop Report (2004).
4. Toutanji, H. Strength and Durability Performance of Waterless Lunar Concrete. Proceedings: 43rd AIAA Aerospace Sciences Meetings, Reno, NV, January 2004.
5. Bodiford, M.P.; Burks, K.H.; Ethridge, E.; Tucker, D. Lunar Contour Crafting: A Novel Technique for ISRU Based Habitat Development. American Institute of Aeronautics and Astronautics Conference, January 2005.
6. Khoshnevis, B.; Carlson, A.; Leach, N.; Thangavelu, M. Contour Crafting Simulation Plan for Lunar Settlement Infrastructure Buildup, ASCE Earth and Space Conference, Pasadena, 15–18 April 2012.

5

The Human Element

Credit: public domain

SELECTION

On 29 April 1961, Leonid Rogozov was writhing in pain due to what he had diagnosed as a suppurative appendix. Rogozov was certain about the diagnosis because he was a physician who also happened to be a crewmember on the sixth Soviet Antarctic Expedition at Novolazarevskaya Station. The problem was that he was the only physician on the base

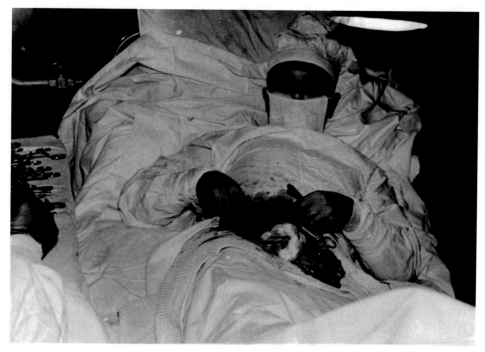

5.1 Leonid Rogozov performs an auto-appendectomy. Imagine trying to do this zero gravity or even a reduced-gravity environment. Credit: British Medical Journal

and the location[1] of the base was … well, Rogozov might as well have been on the Moon. After spending another night in mind-numbing pain, Rogozov decided there was only one option: he would have to perform the operation himself. Which is exactly what he did (Figure 5.1). With some assistance from the expedition's chief engineer and meteorologist, Rogozov (see sidebar) performed the operation in a semi-sitting position. The procedure lasted nearly two hours and, for those of you who are interested in the details of an auto-appendectomy (the medical term for what Rogozov did), an account of the operation featured in the December 2009 edition of the *British Medical Journal* (Rogozov also published a short note of his ordeal in the 1962 *Soviet Antarctic Expedition Information Bulletin*, No. 37, pages 42–4). On his return to the Soviet Union, Rogozov was feted as a national hero (he was awarded the Order of the Red Banner of Labour) and his story was used as propaganda thanks in part to the parallels between Rogozov and Yuri Gagarin (who had made the first manned spaceflight 18 days before Rogozov's operation). Incidentally, Rogozoz never returned to the Antarctic. Can't say I blame him.

---

[1] Novolazarevskaya Station was located at 70°S 11°E in the Shirmacher Oasis in Queen Maud Land: this was more than 80 kilometers from the ice edge.

Excerpt from Rogozov's journal entry

"I worked without gloves. It was hard to see. The mirror helps, but it also hinders – after all, it's showing things backwards. I work mainly by touch. The bleeding is quite heavy, but I take my time – I try to work surely. Opening the peritoneum, I injured the blind gut and had to sew it up. Suddenly it flashed through my mind: there are more injuries here and I didn't notice them ... I grow weaker and weaker, my head starts to spin. Every 4–5 minutes I rest for 20–25 seconds. Finally, here it is, the cursed appendage! With horror I notice the dark stain at its base. That means just a day longer and it would have burst and ...."

"At the worst moment of removing the appendix I flagged: my heart seized up and noticeably slowed; my hands felt like rubber. Well, I thought, it's going to end badly. And all that was left was removing the appendix .... And then I realised that, basically, I was already saved."

So what has the subject of appendectomies got to do with the selection of future lunar crews? Well, since Rogozov's experience, appendectomies have become mandatory for many Antarctic winter-over personnel and it will almost certainly be compulsory for those venturing back to the Moon because the medical capabilities of a lunar base won't be much better than those available to Rogozov more than 50 years ago. A crewmember unfortunate to suffer appendicitis will suffer a range of symptoms, including fever, nausea, pain when walking, diarrhea, and intense pain. Diagnosis is usually by means of computed tomography (CT), but this won't be available on the Moon and evacuation is out of the question because, by the time the crewmember arrives back on Earth, there is a very good chance that the appendix will have ruptured and the crewmember will be dead. That's not a good mission scenario is it? So, one of the first stops in the path to selection as a lunar crewmember will be an appointment with the surgeon for pre-emptive surgery.

Pre-emptive surgery

Our prospective lunar crewmember will first be administered antibiotics such as Unasyn and Flagyl (metronidazole) or perhaps Zosyn (pipercillin-tazobactam), after which a McBurney incision will be made: this is a cut that runs diagonally across the lower right part of the stomach. Next, skin bleeding will be controlled and another small incision will be made to open up the muscle fibers before retracting the muscle belly of the external oblique using a clamp so the surgeon can access the peritoneum (the lining of the abdominal cavity). The peritoneum will then be grabbed by forceps and a small incision will be made to expose the cecal area so the surgeon can inspect the appendix to make sure there is no rupture. The appendix will then be removed and the mesoappendix will be divided by using Kelly clamps and tied off. Once the appendix has been isolated, two ties will be placed on the area of the appendix to be removed and the appendix will be cut out. The abdominal cavity will be flushed with saline solution and antibiotics, and the edges of the peritoneum will be stitched using Vicryl suture. If all goes well, our potential lunar crewmember will be allowed to leave the hospital one or two days later. But, before leaving the hospital, they will check in with the surgeon's office to schedule their next appointment – gallbladder removal, or cholecystectomy.

Why remove the gallbladder? Gallstones is the answer. Gallstones form in the gallbladder (Figure 5.2) and the best way of diagnosing the condition is to use transabdominal ultrasonography or endoscopic ultrasonography. This option will definitely not be available on the lunar surface anytime soon, so we're left with the pre-emptive surgery option again: cholecystectomy. That's because the most common symptom of this condition is biliary colic, and this is something no crewmember wants to deal with on the surface of the Moon. Biliary colic affects 80% of those diagnosed with gallstones. The symptoms are caused when the bile ducts become blocked by a gallstone. Fluid begins to accumulate behind the obstruction, which causes the gallbladder to distend. The result is blinding pain that may last for only 15 minutes or may go on for five hours or longer. It's a mission scenario that the crew can do without, which is why our prospective astronaut now faces his/her second pre-emptive surgery procedure.

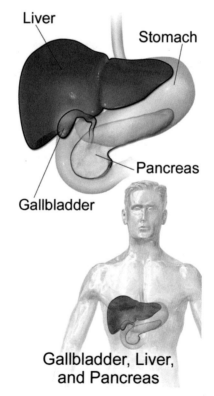

5.2 Gallbladder. Credit: www.blausen.com

After administering local anesthetic (lidocaine is usually used), the surgeon will use the minimally invasive laparoscopic Hasson cannula technique to access the stomach and a small port will be placed through the epigastric wall into the peritoneal cavity so the surgeon can use a laparoscopic camera to see what he/she is doing. A pair of forceps will be placed through the port and the gallbladder will be grasped and retracted. Blunt dissection is then used to free the cystic duct from the surrounding tissue and two clips will be placed on the side of the cystic duct that remains and the other towards the gallbladder. The cystic duct will then be incised and dissected, and the forceps repositioned to move the gallbladder and retracted before complete removal from the liver bed. The Hasson cannula will then be used to irrigate the site and the ports will be removed before closing the incision. After the surgery, our lunar crewmember candidate will stay in hospital for another day or two before returning to their normal activities in a week or so, but not before checking in with the surgeon's office again to make an appointment for one final procedure: sterilization if the crewmember is a woman and sperm banking if the crewmember happens to be male. Here's why.

STERILIZATION AND RELATED ISSUES

All that high-energy radiation in deep space will almost certainly guarantee that Moon-bound or Mars-bound astronauts will be rendered sterile during the course of their missions. Call it an occupational hazard. But, even though all that radiation will almost certainly preclude a pregnancy, it's better to be safe than sorry because the specter of a pregnancy on the surface of the Moon doesn't bear thinking about. That's because the genetic material that develops in a fertilized embryo would be very easily damaged by the type of radiation bombarding astronauts during their deep-space missions. We understand the hazards of ionizing radiation on the reproductive cycle from observing the effects in those who have survived atomic bomb blasts and those who have undergone radiotherapy for cancer. One of the effects of most concern to a prospective off-world parent is the process of cell division. During gestation, cells differentiate very quickly and any damage inflicted on a cell ordained to become a vital organ could be amplified by a radiation burst. The potential outcome could be a baby born with mental deficits. Not that any pregnancy on the Moon would be brought to term because the very first action on the announcement of such an event would be the crew medical officer breaking out the abortifacient. So, to avoid such a dire scenario, male crewmembers will be offered the option of sperm banking in the event they want to have the option of being a father after the mission and females will be whisked off for sterilization.

The most likely procedure to be used will be tubal ligation. This is also known as "having your tubes tied" and is a procedure that closes the fallopian tubes which prevents sperm from entering and fertilizing an egg. The procedure is 99.5% effective and is usually performed as out-patient surgery. Some crewmembers may opt to have their eggs frozen before the procedure if they plan of having a family after their tour of lunar duty.

GENETIC TESTING

Being sterilized and having appendices and gallbladders removed will be a standard pre-selection requirement for prospective lunar astronauts, but what else can this group do to improve their chances of selection? Well, it will definitely help their application if they happen to be radiation-resistant, have high bone density, and aren't susceptible to vision impairment. How will they know this? Genetic testing of course! We know that astronauts bound for the lunar surface will be at high risk of all sorts of ailments, including weakened bones, increased risk of cancer, and cataracts. So, to reduce these risks, it makes sense to select only those crewmembers who have a natural resistance to these problems. Let's take an example. About half of all astronauts suffer back pain during their missions and this pain is usually treated with a mix of exercise and painkillers, with codeine being the pharmaceutical intervention measure of choice. But there are some people who, because of a variant in the liver gene *CYP2D6*, metabolize drugs such as codeine much too quickly. This could cause problems because people with this genetic mutation could easily overdose. But, if astronauts are genetically screened, then those with this mutation could be given lower doses. And metabolism is just one of dozens and dozens of genetic mutations and abnormalities, some of which have been identified thanks to Omics, a blend of genomics, proteomics, transcriptomics, and metabolomics. By applying Omics methodologies, it will be possible to identify the genetic risk profile of each crewmember and thereby create a personalized set of countermeasures. The system will also be able to screen out those potential crewmembers whose bone density is too low and whose radiation sensitivity is too high. To begin this Omics profiling, crewmembers will need to subject to genetic testing, which will include karotyping, molecular studies, and blood sampling.

GETTING ALONG

Although the first missions to the Moon may stay on the surface for only weeks at a time, follow-on missions will stay longer, perhaps months. Talk of long-duration missions to beyond Earth orbit invariably gets psychologists talking of how such missions will push the limits of teamwork and what a tough time astronauts will have surviving in a small base in an isolated and harsh environment (Figure 5.3). These same psychologists argue that teamwork problems could jeopardize the missions and then use this argument to justify boondoggles such as Mars500 (see sidebar) (Figure 5.4). Some researchers and social scientists have even gone as far as to devise badge-like devices that measure biometric data (heart rate, body motion) and how much time crewmembers spend face to face, and identifies when a crewmember *gets loud*, whatever that means. The device, which is best described as a psychosocial sensing badge, is designed to be worn by astronauts to provide feedback about behavior and monitor such variables as vocal patterns and vocal pitch in an effort to determine whether a conversation is friendly or acrimonious. The researchers who dreamt up this gadget argue that such monitoring and self-regulation will be a key element in any long-duration mission beyond Earth orbit because crews may succumb to dysphoria, profound isolation, and homesickness. Of course, crews will feel a little lonely once in a while, and of course they may feel a little out of sorts, but anyone who thinks a

5.3 Ernest Shackleton, one of the greatest explorers of all time. Credit: public domain

5.4 Mars500. Credit: ESA

116 The Human Element

psychosocial sensing badge is required by long-duration crews has a very flimsy knowledge of human exploration (see Figure 5.3). So, instead of wasting time whining about what a rough time long-duration astronauts will have, these researchers should head down to their local library and read a few accounts from the "Heroic Age of Antarctic Exploration." Let's take a short trip down memory lane for those who are unfamiliar with the first group of explorers who had the right stuff.

Mars500: an experiment in how to fritter away US$15 million

Manned spaceflight is littered with examples of money being wasted on programs that went nowhere. Think of the Constellation program on which was spent US$9 billion in addition to another US$2 billion for closing costs. Or take the X-38 program that cost US$150 million before being canned. While both the Constellation and X-38 programs were worthwhile, albeit expensive, ventures, the same can't be said of the Mars500 boondoggle, a pretend spaceship/isolation facility located at the Russian Institute for Biomedical Problems in Moscow. Here, between June 2010 and November 2011, a group of volunteer astronauts simulated a mission to Mars. It was an elaborate – and ultimately futile – mission that was supposed to help psychologists understand how humans responded to the challenges of isolation. I say futile because, if the researchers had bothered to walk down to their local library and searched under "polar exploration," they would have found all the answers they needed. So, while the mission was supposed to have revealed "unique" and important data about how humans cope under adversity, in reality it did nothing of the sort. The simulation wasn't even that realistic because, if there had been a serious problem, then any one of the volunteers could have bailed. But that won't be the case on a lunar base or on a mission to Mars. And it wasn't the case during the "Heroic Age of Antarctic Exploration."

Uncharted territory

"The horror of the nineteen days it took us to travel from Cape Evans to Cape Crozier would have to be re-experienced to be appreciated; and any one would be a fool who went again: it is not possible to describe it .... It was the darkness that did it. I don't believe minus seventy temperatures would be bad in daylight, not comparatively bad, when you could see where you were going, where you were stepping, where the sledge straps were, the cooker, the primus, the food; could see your footsteps lately trodden deep into the soft snow that you might find your way back to the rest of your load; could see the lashings of the food bags; could read a compass without striking three or four different boxes to find one dry match; could read your watch to see if the blissful moment of getting out of your bag was come without groping in the snow all about; when it would not take you five minutes to lash up the door of the tent, and five hours to get started in the morning."

Excerpt from Apsley Cherry-Garrard's Worst Journey in the World *(Picador, 2001)*

Cherry-Garrard's book routinely features in the top 10 lists of the best exploration books ever written, and with good reason. During their epic five-week journey to collect penguin eggs, Cherry-Garrard and his two companions, Dr. "Bill" Wilson and Lieutenant "Birdie" Bowers, endured conditions that were murderously inhospitable. The brutal, unrelenting cold meant even basic tasks became exercises in torture. For example, their sleeping bags, which were made of reindeer skin, became soaked with condensation and then turned rigid with ice. This was also the fate of their state-of-the art expedition footwear, which comprised felt boots called *finnesko*: no Goretex or North Face gear for these guys. Nor did they have energy gels or freeze-dried high-carbohydrate instant meals: instead, Cherry-Garrard and his teammates ate mostly pemmican, butter, biscuits, and tea.

"When we got into our sleeping-bags, if we were fortunate, we became warm enough during the night to thaw the ice: part remained in our clothes, part passed into the skins of the bags...and soon both were sheets of armour-plate. As for our breath, in the daytime it did nothing worse than cover the lower parts of our faces with ice and solder our balaclavas tightly to our heads."

Excerpt from Apsley Cherry-Garrard's Worst Journey in the World

During their trip to chart one of the few remaining blank areas on the map and to gather penguin eggs, Cherry-Garrard, Wilson, and Bowers endured surreal levels of deprivation, hardship, and danger. Gale force 10 winds blew constantly, fingers became frostbitten, and they suffered extreme exposure when their tent blew away.

"Now in darkness it was more complicated. From 11 a.m. to 3 p.m. there was enough light to see the big holes made by our feet, and we took on one sledge, trudged back in our tracks, and brought on the second ... and even the single sledges were very hard pulling. When we lunched the temperature was –61°."

Excerpt from Apsley Cherry-Garrard's Worst Journey in the World

As you read the book, you can't help but shake your head, thinking it can't possibly get any worse. But it does. Winds that blow at 160 kilometers per hour, snow-blindness, and compasses that go haywire because of the group's proximity to the magnetic pole. But this group not only achieved their mission objective (they returned with three penguin eggs), they also survived. And these guys didn't have the benefit of tens of millions of dollars of training. They didn't spend weeks locked inside some simulator to see whether they could survive the real thing. They didn't have to listen to shrinks agonizing about what a terrible ordeal such a journey would be. No, these truly remarkable men were assigned a task and got on with the job. Those attempting to justify Mars500 and similar boondoggles argue such simulations are necessary because researchers don't know how humans will be affected over long periods of time. Of course they don't know because they haven't bothered to read from the rich annals of Antarctic exploration! These same researchers who try

118 The Human Element

to dream up ways to squander millions of dollars on unproductive simulations argue that good teamwork will be essential on long-duration missions and that the only way to identify problems is to lock up volunteers in tin cans. Rubbish! We know damn well that humans can function well as a team under the most appalling circumstances. Ever heard of Roald Amundsen? Fridtjof Nansen? Douglas Mawson? Sir Ernest Shackleton? The Mars500/tin-can researchers obviously haven't, so here's a reminder.

In December 1914, Ernest Shackleton set sail on the *Endurance* together with a 27-man crew. His goal: to accomplish the first crossing of the Antarctic continent. Not one to pull any punches, the great explorer was supposed to have posted the following recruitment ad: "Men wanted for hazardous journey. Small wages. Bitter cold. Long months in complete darkness. Constant danger. Safe return doubtful. Honour and recognition in case of success." Incidentally, Shackleton received 5,000 applications, which is more than 1,000 more than the number that applied for NASA's 2009 Astronaut Class. Oh how times have changed! But let's get back to Shackleton (Figure 5.5). Shortly after arriving on the Antarctic continent, the *Endurance* became trapped by ice (Figure 5.6) and, for 10 months, the ship drifted before pressure crushed the ship, leaving the crew with little food, clothing, or shelter.

5.5 Men with the 'Right Stuff'. Nimrod Expedition South Pole Party. Left to right: Wild, Shackleton, Marshall, and Adams. They don't make explorers like this anymore! Credit: public domain

5.6 The *Endurance* being crushed by the ice. Credit: public domain

After camping on ice floes for five months, the crew finally drifted to the northern edge of the ice pack and sighted open leads of water. The crew set sail in three small lifeboats and landed on Elephant Island. It was the first time the crew had set foot on land in 497 days. But Elephant Island was far from the shipping lanes and rescue was unlikely. So Shackleton took five crew with him in the lifeboat named the *James Caird* and struck out for South Georgia Island on which a whaling station was located. After 17 days crossing some of the roughest seas on the planet, Shackleton and his men landed on South Georgia but then had to cross 42 kilometers of mountainous terrain (which wasn't even mapped until the 1950s) to reach the whaling station. Frostbitten and starved after their 21-month epic, Shackleton and his men reached the whaling station and then returned to rescue the remainder of his crew on Elephant Island. Not a single crewmember was lost thanks to teamwork and extraordinary leadership. And remember, most of Shackleton's crew were recruited right off the dock. They didn't spend years in pre-mission training and they didn't have access to Facebook or iPhones, or the internet. They suffered incredible privation and frostbite, and subsisted for the most part on a mix of penguin meat, pemmican, and hoosh. I wonder what Shackleton would have made of the Mars500 experiment and all the other analog missions? There are so many parallels between long-duration spaceflight and polar exploration that you could write a book on the subject, which is what I did. It's called *Survival and Sacrifice in Mars Exploration: What We Know from Polar Expeditions*.[2]

MARTIAN ANALOG MISSION

But let's not discount the usefulness of analog missions entirely. Let's just tweak the idea a little and devise a mission with more relevance to the long-term objective, which is a manned mission to Mars. Until the Variable Specific Impulse Magnetoplasma Rocket (VASIMR) is operational, a Mars mission will take about 30 months, so the agencies and commercial companies should put their heads together and deploy a 30-month mission to the Moon. The mission would begin with a seven-month period orbiting the Moon, a surface stay of about 16 months, followed by another seven-month period orbiting the Moon. Why not orbit Earth, you may ask? Well, we want to make sure the crew is exposed to as much deep-space radiation as possible: just like they would on a regular Mars mission. Think how valuable such a mission would be. Not only would it build on the data acquired during the many six-month missions to the International Space Station (ISS) and Mir, it would also provide new insights into how humans adjust to really, *really* long-duration spaceflight. Following the mission, the data derived from the crew would be shared among the participating agencies and/or companies, and that data could be used to reduce mission cost and refine procedures for an actual mission to Mars. We know that six-month missions to the ISS are not the best model – not even close – for a trip to Mars so why not

---

[2] *Survival and Sacrifice in Mars Exploration: What We Know from Polar Expeditions* was published by Springer Praxis Books in March 2015. ISBN-13: 978-3319124476.

5.7 Survival training. Credit: CSA

strike out, be bold, and conduct a high-fidelity field test? Something that even Shackleton would appreciate perhaps? Here's how it might work.

After being sterilized and having their appendices and gallbladders removed, our four crewmembers (one flight surgeon, one engineer, one scientist, and one pilot), representing the US, Europe, and Russia, would spend a year training for the mission to ensure they are proficient in the multitude of tasks required of an interplanetary crew. This training will include systems training, survival training (Figure 5.7), lectures on life-support systems, and instruction on the myriad scientific protocols the crew will be expected to complete. The scientific investigations will focus on seven fields of study, the first being functional studies to assess how physiological factors change after 30 months in reduced gravity.

The second field of study will be behavioral health, which will examine such factors as sleep and neurocognitive performance, although there will be no studies examining the effects of isolation because we know how humans perform in isolated environment for prolonged periods of time. The third field of study will examine the extent of visual impairment on the crew. To date, NASA only has data on crewmembers who have spent six months on orbit. How much worse will the visual impairment (Figure 5.8) problem be after 30 months? And don't forget our crew will be exposed to deep-space radiation and therefore there is a good chance some or all the crew will develop cataracts. How will they deal with that problem? No abort or rescue capability on this mission remember – just like there won't be any rescue on a real mission to Mars.

122   **The Human Element**

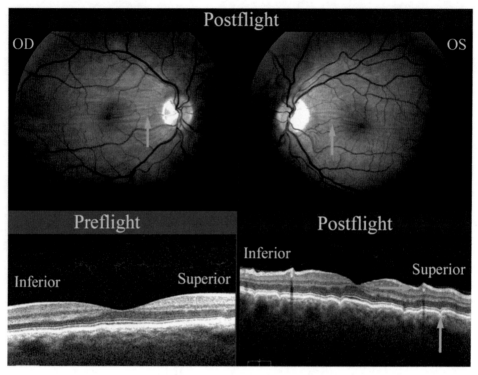

5.8   Visual impairment intracranial pressure (VIIP). Credit: NASA

The fourth field of study will be metabolic investigations aimed at defining the extent of oxidative stress during what will be a very long mission, while the fifth series of investigations will examine physical performance. This will be particularly interesting because we know some astronauts lose as much as 1% of their bone mass per month so there is a good chance some of the crew could lose up to a third of their bone mass by the time they return to Earth. Will they be able to conduct all their mission tasks on the surface of the Moon without risk of fracture? We won't know until the crew embarks on this mission. The sixth field of study will be human factors that will examine how this pioneering group interacts with their environment en route to the Moon and while they are working on the lunar surface. And finally, the seventh area of investigation: rehabilitation post flight. Can this crew be rehabilitated? Will they even be able to walk again? No need for any post-mission psychological studies because this crew will have spent 30 months in the proverbial lap of luxury compared to the travails of Shackleton, Cherry-Garrard, and company. We know more than enough about psychological resilience to know this spacefaring group of very highly trained individuals will have no trouble dealing with 30 months of being hammered by deep-space radiation or losing large chunks of their bone mass. And, on the subject of risk, let's not pull any punches here: this mission will lie on the very ragged edge of what is physiologically survivable, which means this crew will receive some rather unique training, especially in the arena of bioethics. What will they do if a crewmember dies? What are the procedures for burial in space? On the surface of the Moon? What is the course of action to deal with a terminally ill crewmember? How does one go about euthanizing a fellow astronaut? The answers can be found in the field of space bioethics.

> "From its inception, space exploration has pushed the boundaries and risked the lives and health of astronauts. Determining where those boundaries lie and when to push the limits is complex. NASA will continue to face decisions as technologies improve and longer and farther spaceflights become feasible. Our report builds upon NASA's work and compiles the ethics principles and decision-making framework that should be an integral part of discussions and decisions regarding health standards for long duration and exploration spaceflight."
> *Jeffrey Kahn, Chair of Committee, Health Standards for Long Duration and Exploration Spaceflight*

DEATH SIMS

For the history of spaceflight, the philosophy of orbit-based medicine is to treat what is treatable and to evacuate in the case of a serious injury or illness. This philosophy will not work for long-duration spaceflight beyond low Earth orbit because crews will be at the very least several days from a return to Earth. So, to help crews deal with potentially mission-threatening medical events, it will be necessary to recalibrate current values and priorities that guide medical decision-making. It will also be necessary to make some very tough calls about how much medical equipment to take along to the Moon and to Mars, and also to balance likelihoods and needs. Once those decisions have been made, crews will need to be furnished with a decision framework that assists them in making decisions about medical resource allocation and what to do in the event of more serious medical contingencies. For example, imagine a scenario in which a crewmember sustains a serious head injury while working on the lunar base. This crewmember isn't expected to live more than four of five days, so what does the crew do? Evacuation isn't an option and the injured crewmember is using up life-support consumables that could lengthen the mission. Also, caring for the stricken crewmember removes two other crewmembers from their duties and the mission is short-staffed as it is with a crew of only four. Euthanization is probably the best option, but what does the crew do with the body? Store it in the base until the next crew arrives in four months? Unlikely. Burial on the lunar surface perhaps? But what are the procedures for burying a crewmate in space? These, and many other bioethical policies, can be found in the *Interplanetary Bioethics Manual*. This isn't published by NASA, but it can be found as an appendix in my book, *Interplanetary Outpost*.[3]

---

[3] *Interplanetary Outpost: The Human and Technological Challenges of Exploring the Outer Planets* was published by Springer Praxis Books in 2012. ISBN-13: 978-1441997470.

Just because NASA – or any other space agency for that matter – hasn't published a bioethics manual doesn't mean the agency is avoiding the problem. For from it. Crews routinely participate in a training exercise known as the "death sim," which gets astronauts thinking about what they might have to do if one or more of their crewmates die. The death sims usually involve the whole crew who are presented with a hypothetical mission scenario in which a crewmember has died. As the scenario plays out, the crew are presented with questions such as what should be done with the body and which next of kin should be informed. It's a pretty grim exercise but it gets crews thinking about the "what ifs."

6

Regulation

Credit: NASA

Over the years, there has been a lot of talk about establishing a base on the Moon for the purpose of extracting helium-3 (He-3), a rare isotope of helium that is believed to exist in large quantities on the lunar surface. Theoretically, He-3 could serve as a fuel for thermonuclear fusion reactors, thereby providing a practically limitless energy source. For example, it is estimated that just 40 tonnes of liquefied He-3 would provide enough energy to meet the electrical demands of the US for a year! While a question mark remains over the viability of He-3-based fusion energy, a bigger question mark looms over the

126 Regulation

6.1 A Shackleton Energy Company (SEC) vehicle. Credit: SEC

exploitation of the resource on the Moon because there is no international agreement on whether a nation or commercial entity (Figure 6.1) is allowed to acquire title to lunar resources. The 1967 Outer Space Treaty, developed by the UN, provides plenty of guidelines, but no regulation about the exploitation of lunar resources. The 1979 Moon Treaty, another UN-sponsored policy document, does address the resource exploitation question, but it has only been signed by a few countries, none of which is a major space power. So what are the prospects of mining He-3 and conducting other commercial activities on the lunar surface?

MOON TREATY

We'll start with the aforementioned Moon Treaty, but first a primer on international law. International law is a rather unique element in the legal arena because it can be applied as custom and treaty. For instance, naval vessels belong to the country of origin unless they are abandoned by that country. This means that, if one of Canada's submarines (Figure 6.2) were to sink, it would not be subject to the international salvage laws no matter where that sub sank. This is an example of customary law as it applies to maritime law and customary law also applies in space: when Sputnik was launched, customary law ensured the satellite enjoyed free passage when transiting over sovereign nations.

6.2 HMCS *Victoria*. Credit: Canadian Forces

Now when it comes to international law, the situation is a little different because of something known as "treaty ratification." A nation that ratifies a treaty is legally bound by what is stated in that treaty and that means legal force can be applied to ensure the terms of that treaty are complied with, but it's not quite that simple because the weight given to a treaty will depend on whether a nation signed or acceded to that treaty. A nation that accedes to a treaty will be legally bound by what is stated in that treaty whereas a nation that only signs a treaty is not legally bound because a signature only indicates that nation's intent to look more closely at the treaty before making a commitment to ratify it, if in fact they do.

So where does the Moon Treaty fit into all of this legalese? The Moon Treaty was adopted and opened for signature by the UN in 1979, although it did not come into force until 1984. To date, the Moon Treaty has been ratified by six countries and four countries have signed it. Russia, China, and the US have neither signed, acceded to, nor ratified the treaty, so what is the purpose of this document? Well, the Moon Treaty provides that the lunar environment should not be disrupted and that it should only be used for peaceful purposes. It also provides that the UN should be informed of the purpose and location of any base of habitat on the lunar surface and that no part of the lunar surface can be owned by a company, a person, or any other entity unless that entity happens to be an international organization or a government. Perhaps the most contentious part of the treaty is the section dealing with natural resources. The Moon Treaty provides that the resources on the Moon cannot be exploited unless an international regime is created to oversee that exploitation.

No definition of what "resources" are or what an "international regime" constitutes is specified in the treaty but it is safe to assume that He-3 and water ice would be classed as resources. There is a precedent of sorts for the notion of an international regime and it was suggested as part of the 1994 Agreement of the Law of the Sea Convention. In this case, it was proposed that an international regime – The Enterprise – would supervise the mining of minerals in the oceans (Figure 6.3). It was proposed that developed nations and commercial entities would be supervised by The Enterprise and that a percentage of the wealth derived from the exploitation of minerals would be given to The Enterprise to be distributed among developing countries. The Enterprise also proposed that developed nations transfer their technology to The Enterprise so this technology could be used by developing nations to also extract mineral wealth from the oceans.

As you can imagine, a similar international regime wouldn't go down very well with the nations and/or commercial entities planning on extracting resources from the Moon. Can you imagine the Chinese relinquishing a percentage of their hard-gained lunar resources or asking the US to surrender technology to the Chinese? It isn't going to happen. But that is the common heritage view suggested by the Moon Treaty. But does this even matter because the major spacefaring nations are not parties to the Moon Treaty, which surely means the treaty is a failure? Perhaps, but the treaty, even though it is not technically binding on the non-parties, is still technically valid in international law. From

6.3 Mining the oceans. Credit: Wiki

the perspective of the US, the Moon Treaty has been on a backburner for a while because President Reagan indicated that his administration was against ratifying the treaty and, since then, the US has taken no official position regarding the validity of the treaty. So where does this leave the US legally? Can Bigelow Aerospace just go ahead and set up base on the Moon and start mining He-3 without fear of legal reprisals? That would be one option, but imagine what might happen if that occurred. Perhaps Russia and China, who are also gunning for the Moon, would decide to ratify the Moon Treaty, thereby putting international pressure on the US to comply? It's an unlikely game-changer, but possible nevertheless. After all, these are two space powers, and their being non-parties to the treaty is a significant argument for the treaty being a failed piece of legislation. But if, all of a sudden, these countries were to ratify the treaty, then perhaps that would revive the reputation of the treaty and force the US to recalibrate their lunar exploitation. What might happen then? It is possible the Russians and the Chinese could exert diplomatic pressure on the US, with the result that the US signs the treaty. Such an action wouldn't require the US to ratify the treaty and it would actually be to their benefit because now they could seek ways to amend the treaty and perhaps request that some parts be nullified. It would probably be a better strategic move than simply turning its back on the treaty.

Now you may be thinking that, because the three major space powers haven't signed the treaty, this must surely mean that this is a piece of legislation that has no teeth, so why don't the US or China or Russia simply go it alone? The reason is customary law, but that can't truly be tested until the exploitation of lunar resources becomes an actuality.

> "The national regulatory framework, in its present form, is ill-equipped to enable the U.S. government to fulfill its obligations."
> *Excerpt from FAA letter written by George Nield, associate administrator for the FAA's Office of Commercial Transportation in response to query by Bigelow Aerospace*

PROPERTY RIGHTS

In December 2014, Bigelow Aerospace received a response from the Federal Aviation Administration's (FAA) Office of Commercial Space Transportation regarding the matter of a payload review query regarding the staking of claims on lunar territory. In what was widely seen as a preliminary step by the US government to encourage exploitation of lunar resources, the FAA informed Bigelow Aerospace that FAA officials "recognize the private sector's need to protect its assets and personnel on the moon or on other celestial bodies." This didn't mean the FAA was authorized to give Bigelow Aerospace a license to set up a mining operation on the Moon, but it did document a proposal made by an American company to request a launch license to pursue that objective sometime in the future. The response also recognized Bigelow Aerospace's need to protect its assets and personnel on the Moon. So, in essence, what the FAA's response meant was that, if Bigelow Aerospace or Shackleton Energy Company (SEC) (Figure 6.4) were to set up shop on the Moon, then no one else would be authorized to set up base at the same location. Makes sense.

130 **Regulation**

6.4 Jim Keravala of Shackleton Energy Company (SEC). Credit: SEC

The reason Bigelow Aerospace's enquiry was necessary was because commercial space technology is moving at a faster pace than the legal framework that is needed to govern that technology. This shouldn't be too surprising because developing legal frameworks is difficult enough on Earth and the few treaties that are in place to govern on-orbit and beyond-orbit activities are vague at best. After all, when the Moon Treaty and Outer Space Treaty came into being, the notion of asteroid mining and tourist trips to the Moon existed only in the pages of science fiction. So where does the law stand with regard to property rights on the Moon? Will it be possible for Bigelow Aerospace or SEC to adopt the Frontier Paradigm perhaps, or maybe they will simply occupy a spot on the Moon and claim corporate possession (*corpus possidendi* in legalese). Probably not, but the appropriation of material is a different kettle of fish. In this case, Bigelow Aerospace or SEC could claim enterprise rights that would permit them to exploit the Moon's resources. Which is why Bigelow's letter to the FAA was an important step in the direction of resolving the issue of property rights because, without property rights, it may prove difficult to persuade investors to sink money into this endeavor. Let's face it, lunar resources will be exploited; it's just a question of when and by whom. There are some space entrepreneurs who suggest the Homesteading Act could work as a model to kick-start resource extraction on the lunar surface. The Homesteading Act helped develop the American West by allowing people to apply for a land grant for the piece of land where they lived and worked. It might be a good model but it couldn't be implemented under the Outer Space Treaty that prohibits appropriation. And developing the lunar surface is very, *very* different from pioneering the Wild West. For one thing, the legal and geopolitical environment is very different today than it was in the nineteenth century. For instance, the Homesteading Act was devised to consolidate American power and the eligibility requirements for being granted a land grant was to be over 21 years old and promise not to take up arms against the government. In sharp contrast, setting up shop on the Moon requires bucketfuls of money and a spacecraft.

"As far as title goes, it's a gray area. And from a risk perspective, lack of clarity means it doesn't exist."
Lawyer and space-law expert Timothy Nelson in an interview with SPACE.com

Going even further back in time, the Romans established two concepts for owning resources: *res nullius* and *res communis*. The former applied to resources such as gold in the Wild West, which meant that anyone who could find and mine the resource owned it. So, say you were an off-world mining company and you wanted to snag an asteroid (Figure 6.5). Under *res nullius*, you could simply lay claim to the asteroid and start mining it for resources. No legal hurdles.

But the Moon seems to be more aligned with the concept of *res communis*, or communal property. There are many people who object to the notion of corporations owning parts of the ocean and these people also have a problem with the idea that anyone can own part of the Moon, arguing that our planetary neighbor should be common to all mankind. Send scientists and explorers by all means, but don't start strip mining the surface. It's a noble vision, but noble visions rarely prevail. What is likely to prevail will be a sustainable approach to the extraction of lunar resources which will allow companies to profit from the mining of materials but will prevent any entity from owning any part of the Moon.

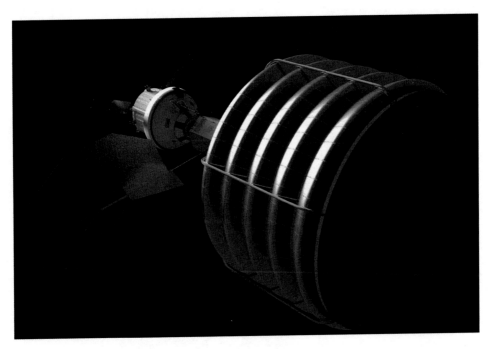

6.5 NASA's asteroid redirect mission. Credit: NASA

SALVAGE

It is December 2027 and the Chinese have landed on the Moon. It is the first time humans have stepped on the lunar surface in exactly 55 years. On the second mission day, two taikonauts wander over to Tranquillity Base, pull up the American flag, and return to their lander. Three weeks later, the flag is on display in Beijing's Museum of Science and Technology. NASA is understandably upset and Sino-American relations take a nosedive. Would this be allowed? To answer that, it is useful to have a basic understanding of salvage law and how this law may be applied to those operating on the Moon.

The word "salvage" conjures up images of saving ships and/or their cargo, which is why salvage will likely be a prosperous business when lunar surface operations commence. After all, repurposing hardware and recovering equipment save money. Here on Earth, the Maritime and Admiralty Law of the US defines salvage as "the compensation allowed by persons by whose assistance a ship or her cargo has been saved, in whole or in part, from impending peril on the sea or in recovering said property from actual loss as in cases of shipwreck, derelict or recapture." The international version of salvage is stated in the 1989 Brussels Salvage Convention, which specifies the three conditions that are necessary for a legal salvage [1]. The first stipulation is that there must be danger to the property that is to be salvaged, the second is that there must be a voluntary service not owed to that property, and the third stipulation is that the salvage operation must have been at least partially successful in retrieving said property. When it comes to the legal stuff, salvor claims are judged on a variety of factors such as the skill with which the salvage was carried out, the risk the salvors took in conducting the operation, the speed with which the salvage was performed, and the value of the property used by the salvors. These factors are also taken into consideration when determining the award that is to be paid to the salvor: in most salvage cases, the salvor does not become the owner of the property that has been salved, but has a lien against the property that can be used to pay costs in case the property owner doesn't pay up. The only time a salvor would become the owner of the ship or cargo salved would be in those cases when the ship or cargo in question had been abandoned.

So how does modern-day salvage law relate to activities on the Moon? Well, the lunar surface, like the rest of space, is defined as an international area that exists beyond national boundaries. So, just as nations have jurisdiction and control over their terrestrial ships, so too would they retain jurisdiction and control over their spacecraft and bases on the Moon. And salvage operations may also be relevant to personnel involved in the event of an emergency, which is why it is worth mentioning the 1968 Rescue and Return Agreement at this point. What would happen if an American spacecraft suffered an emergency and crash-landed on the Moon, perhaps on a Chinese area of operations? Would the Chinese render assistance and, if they didn't, who would? A commercial salvage operation perhaps?

On Earth, the regulations are clear because, if a spacecraft were to land on foreign territory, that country is required by international law to take all steps necessary to render assistance. Would this same rule apply to the Moon? And what about the salvage of objects such as the Tranquillity Base flag or one of those Apollo rovers (see sidebar)? Can a nation or entity land on the Moon and bring those back to Earth? To answer this, it is necessary to know how those items of hardware on the Moon are classified. For example, one interpretation of the law is that jurisdiction ceases when the owner can no longer control it, which would mean that all those rovers and other items of exploration hardware on the Moon would be classified as abandoned. Another interpretation – and the one most likely to be abided by – is the view that these pieces of hardware remain under the jurisdiction of whichever nation registered them.

Stealing Apollo

The Apollo astronauts left an awful lot of stuff behind when they left the Moon: more than 100 items at Tranquillity Base alone. Some of the more notable artifacts include the space boots worn by Neil Armstrong and Buzz Aldrin. Obviously, NASA is interested in preserving the Apollo sites and protecting those artifacts, but can they do this legally? Once again, the Outer Space Treaty provides some guidelines, since Article II states that no party to the treaty can assert sovereignty over the Moon, which means NASA cannot enact regulations preventing vehicles from operating around any of the Apollo sites [2]. So, while many would like to designate Tranquillity Base a heritage site, any such move would be a symbolic measure. But Article VI of the Outer Space Treaty also provides that nations retain jurisdiction over their spacecraft, although this doesn't mean the US can regulate what non-governmental spacecraft do on the Moon. So where does that leave NASA when it comes to protecting the Apollo sites? Well, there are some space laws that govern foreign entities, one of which is the 1972 Convention on International Liability for Damage Caused by Space Objects. This space law tool, known as the Liability Convention, states that the country launching a vehicle is liable for damage to another nation's property in space "if the damage is due to its fault or the fault of persons for whom it is responsible." So, if a commercial vehicle landed on top of and destroyed some Apollo artifacts, then the US could invoke the Liability Convention and claim damages. Also, under the provisions of the Outer Space Treaty, all parties must supervise their activities with "due regard to the corresponding interests" of all other parties. The treaty also stipulates that all parties consult one another in an effort to prevent interference. So, if a commercial or foreign entity damaged an Apollo site, the US could invoke the treaty's provisions and request that the offending nation or entity act responsibly. But, this could only be done if the foreign entity was subject to US jurisdiction and this would only apply if that entity had business registered in the US. And, as far as tort law is concerned, the claim could only be submitted after the damage had been done because there is no legal mechanism for the US to obtain an injunction against an operator that is intent on causing damage or removal of an artifact. So the Chinese, or any other nation interested in displaying Apollo artifacts, could in theory land on the Moon, collect whatever artifacts they're interested in, and argue that the US has abandoned the sites, thereby relinquishing ownership.

The upshot of all this legal ambiguity is that there is little in the way of space tort laws or traditional property laws as applied to the Moon when it comes to protecting lunar sites, whether those sites be deemed of historical interest or otherwise. To adequately protect the Apollo sites and the sites of future commercial operations, it will be necessary for broader international agreement on how to enforce property and retain ownership. To that end, NASA came up with its own set of recommendations that were outlined in a document released in July 2011. The title of the document was "NASA's Recommendations to Space-Faring Entities: How to Protect and Preserve the Historic and Scientific Value of U.S. Government Lunar Artifacts." In the document, the agency cites chapter and verse of myriad legal mechanisms that govern the activities of those exploring the lunar surface and also provides the following rationale for the recommendations outlined in the document:

"RATIONALE: Since the completion of the Apollo lunar surface missions in 1972, no missions have returned to visit these historic sites, leaving them in pristine condition and undisturbed by artificial processes (the sites have changed due to normal space weathering). It is anticipated that future spacecraft will have the technology and their operators will have the interest to visit these sites in the coming years. These visits could impose significant disturbance risks to these sites, thus potentially destroying irreplaceable historic, scientific and educational artifacts and materials. A site may include multiple areas of interest, depending upon the specific mission. For example, the Apollo 11 site can be easily included within a single boundary whereas the Apollo 17 site, with additional mobility provided by the lunar rover, may include multiple boundaries around the landing area as well as around each of the traverse sampling sites."

"NASA's Recommendations to Space-Faring Entities: How to Protect and Preserve the Historic and Scientific Value of U.S. Government Lunar Artifacts," page 7

The document then describes in detail the agency's recommendations for over-flight and near-over-flight of the Apollo heritage sites, citing the danger of creating damaging plumes and ejecta flux on the descent. The document also outlines courses of action in the event of loss of thrust in the vicinity of the heritage sites and defines what it believes to be safe artifact boundaries to prohibit visits by other nations or entities. It also outlines recommendations on how to prevent contamination of the artifacts and designates collision windows. While the intent of the document was to provide recommendations of how to protect the Apollo sites, what NASA has created is a template of guidelines that could be adopted by all nations and entities.

COMMERCIAL OPERATIONS AND RULES OF THE ROAD

Many years ago, when I was training to be a naval officer, one of the many, *many* tests we had to pass was the Rules of the Road (ROR) exam, which basically meant we had to memorize more than 50 pages of legalese and be able to quote chapter and verse of any of the 38 rules (with myriad exemptions). The purpose of the ROR is to maintain the discipline of marine traffic and minimize the risk of an accident and, with a number of commercial operations headed for the Moon over the next couple of decades, perhaps there should be some guidelines on how these entities operate. Here are some suggestions.

We'll begin with the approach path of vehicles landing on the Moon. Obviously, operators working on the lunar surface don't want to have their mining activities affected by rockets landing nearby or on their site, so it will be necessary to establish descent and landing boundaries. Such a boundary would comprise a vertical component of at least five kilometers and a radial distance of three kilometers from the perimeter of operations. These distances would ensure that particles created by a vehicle overflying the site would not affect that site and that any exhaust would not be blown onto the site. The distances are based on mathematical plume modeling of ejecta which predicted that smaller particles could achieve velocities of up to 2,000 meters per second. And, because there is no atmosphere on the Moon, those particles would continue for quite some distance before settling on the surface. And, because lunar soil is highly abrasive, a commercial operator wouldn't take too kindly to another operator's rockets sand-blasting their equipment. Also, in the event of a system failure, the visiting operator would have plenty of space to perform an abort scenario without colliding with any structures belonging to the mining operator. And, on the subject of landing rockets, there would also need to be some guidelines for the use of natural barriers such as craters and ridges to block the plumes and ejecta.

With increasing robotic and human missions, there will come a time when landing activities will need to be coordinated to prevent collisions and also to identify landing windows for sites where the landing terrain may be obscured due to surface operations. And, on the subject of surface operations, there will need to be guidelines for all those rovers scouting the lunar surface. Operators will no doubt establish their own exclusion zones to prevent traffic from entering their area of operations, but what happens to rovers that break down or reach their end of life? If they are abandoned, there is a chance that the extreme lunar environment could cause uncontrollable events such as battery venting and subsequent contamination. And how fast should these rovers be allowed to be driven? The faster the rover's speed, the greater the distance that dust can be thrown up. For example, at four meters per second, a rover would cast debris up to 10 meters but, if that rover accelerated to 18 meters per second, it would cast debris up to 200 meters away. No Formula 1-style driving then! And finally, with so much traffic on the lunar surface, there should also be some provisions to deal with the issue of pollution, such as flying dust from rocket plumes and residual propellants.

POLLUTION

Samples collected by the Apollo astronauts revealed He-3 concentrations in the regolith are as high as 30 parts per billion (ppb). That may not sound like a lot, but He-3 is expensive stuff, with a projected value of around US$40,000 per 25 grams. The problem with He-3 is that it exists in such low concentrations, which means huge areas of rock and soil will need to be processed to isolate the material. Experts reckon that excavating a two-square-kilometer patch of lunar surface down to a depth of three meters would yield about

100 kilograms of He-3. While such an amount would be sufficient to power Oslo *and* Stockholm for a year, all that excavating is likely to leave an unsightly mess on the surface because strip mining is the most efficient means of digging up all that regolith. Here's how it might work. A two-drum dragline would be pulled by cables across the surface to scrape up the regolith before dumping it in a truck. On Earth, discarded material is usually backfilled, but would this happen on the Moon? And what happens to all those gases released by the processing of the regolith? After all, we know that the Apollo missions added about 60 tonnes of gases onto the lunar surface and some of these gases were probably absorbed by the lunar surface before being bound into a solid state. What would be the long-term effects of hundreds of tonnes of gases being released? And what happens if the Chinese are the first to start the mining of He-3? After all, this is a country in which more than a million deaths are attributed to pollution each year and in which eight of the 10 most polluted cities in the world are located. Hardly a beacon for environmental stewardship. So what are spacefaring nations of the world to do to prevent irreversible scarring of the lunar surface once these He-3 mining (and other) operations are up and running? Well, at the very least, nations and commercial entities should be responsible for removing all their equipment and installations at the end of their period of operations and return them to Earth. They should also be required to remove and dispose of all pollutants, minimize space debris in lunar orbit, and, to the greatest extent possible, repair any environmental damage such as scarring of the surface.

REFERENCES

1. Brussels Convention: Convention for the Unification of Certain Rules with Respect to Assistance and Salvage at Sea, 23 September 1910.
2. Outer Space Treaty: Treaty on Principles Governing the Activities of States in the Exploration and Use of Outer Space Including the Moon and Other Celestial Bodies, 27 January 1967, 18 U.S.T.2410, T.I.A.S. No. 6347, 610 U.N.T.S. 205 (entered into force on 10 October 1967).

7

Headwinds and Tailwinds

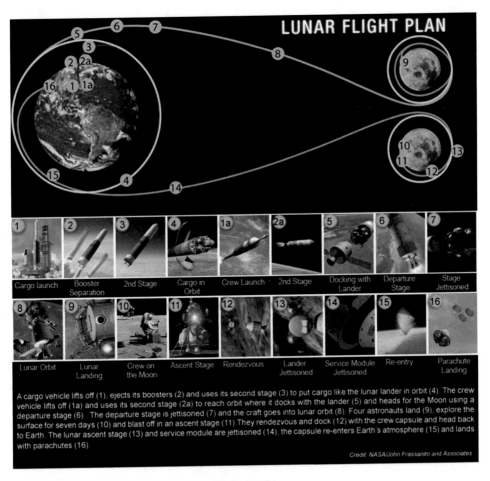

Credit: NASA

138 Headwinds and Tailwinds

7.1 The Constellation program's Altair lunar lander. Credit: NASA

Remember the Constellation program? This venture was hailed as the heir to the aging Shuttle fleet with the bonus that it would return humans to the Moon. As we now know, NASA's new vision was met with almost as much controversy as the program's cancellation. While the crew capsule – Orion – looked like a beefed-up Apollo variant, the scale of Constellation was pretty dramatic, with lots of exploration options thanks to the creation of two rockets: Orion and a new lunar lander (Figure 7.1). The problem was that the program was way over budget and behind schedule, so President Obama decided to create a new path for NASA – one in which there would be increasing reliance on the commercial space sector.

The Constellation program was just one example of the many tailwinds and headwinds that crop up in the manned-spaceflight arena. Hailed first as a tailwind that would see humans return to the Moon, the program's cancellation was rapidly downgraded to a head-tailwind when the planned new boosters and vehicles were consigned to the scrap heap. So, as impetus for a return to the Moon and then Mars takes shape, it is worth considering how these plans can be accelerated and how they can also be derailed.

HEADWINDS

Grounded

Thanks to the excellent film *Gravity*, space debris has had its Hollywood moment. For 90 minutes, Sandra Bullock dodged fast-moving space junk caused by a rogue satellite impacting the International Space Station (ISS). But, while *Gravity* was fictional, the risk of space junk is very real, with more than 23,000 objects being tracked in low Earth orbit

7.2 Space debris. Credit: ESA

(LEO) by the Department of Defense's Joint Functional Component Command for Space (JFCC–Space). And these objects zip around at speeds of eight kilometers per second. Over the years, the ISS and the Shuttles have had to take evasive action to dodge space junk and the 2007 Chinese anti-satellite test and the 2009 collision between a US communications satellite and a Russian satellite haven't helped matters. In fact, the situation gets worse every year, with the result that what was depicted in *Gravity* could actually happen and then all bets to the Moon are off. The cascading series of collisions dramatically portrayed in Alfonso Cuarón's film are based on the Kessler Syndrome[1] – a theory proposed by Donald Kessler in 1978. Kessler, an astrophysicist working for NASA at the time, proposed that a self-sustaining series of cascading collisions of space objects could litter LEO with such a high density of junk that space travel would be too dangerous because of the risk of impact. While the JFCC–Space tracks many of the objects and NASA's Orbital Debris Program Office in Houston tracks debris using radar and data from spacecraft, there is no mechanism in place to actually clean up the mess in LEO. And, every year, the risk of a Kessler Syndrome increases (Figure 7.2).

---

[1] Incidentally, Kessler didn't coin the term. That honor goes to his colleague, John Gabbard, who worked for the North American Aerospace Defense Command (NORAD). He explained Kessler's report to a reporter, who then published the term.

Take the Envisat event, for instance. Envisat was launched by the European Space Agency (ESA) in 2002 to monitor the oceans, atmosphere, and ice caps. At eight tonnes, the satellite was, at the time, the largest Earth-observation satellite ever launched. Then, in 2012, Envisat became the largest inoperable satellite and a significant collision threat since it orbits at an altitude of 790 kilometers – an altitude where there already happens to be a high density of space debris. Already, vehicles have had to take evasive action to prevent colliding with the wayward US$1.8 billion satellite and there is a chance – unlikely, but possible – that the satellite could trigger a Kessler Syndrome event. What can be done? Well, LEO needs to be cleaned up, whether this means using space tugs to de-orbit retired satellites or using drag enhancement devices to drag the same satellites to a lower altitude so they will burn up in the atmosphere. But, while several methods have been proposed, none has been implemented.

Financial crisis

This could be a headwind, depending on which agency you work for. In the 2008 financial crisis, NASA's response was to make cutbacks and ride out the storm whereas ESA actually increased funding for science and research to stimulate investment in innovation. The different approaches reflect the fact that NASA is a government agency whereas ESA is a multinational organization. So what might happen to the return-to-the-Moon program when the next financial crisis hits? If it's a government program we're talking about, then we're probably talking about cutbacks but, if it's a commercial enterprise that is in question, then the outcome may be a little different. That's because politics is a feature of space exploration and government-funded agencies like NASA are subject to political decisions, but a commercial enterprise is a different animal.

INTERNATIONAL LUNAR DECADE

In November 2014, I was fortunate enough to be invited to attend the Next Giant Leap (NGL) conference at Waikoloa on the Big Island of Hawai'i. One of the themes of the conference was how to get back to the Moon sooner rather than later and one of the mechanisms proposed to achieve that goal was international collaboration via a framework known as the International Lunar Decade (ILD). Proposed by members of a group that forms the International Lunar Decade Working Group (ILDWG), the idea is that the ILD will be a global event that aims to fulfill the goals stated in the declaration (see Appendix II). The date for when the ILD will start is 2017, which is the 60th anniversary of the International Geophysical Year. If it kicks off successfully, then the ILD will pursue a series of project objectives that include identifying resource locations and deciding where the next manned mission will land. It will also identify locations where scientific and commercial facilities can be sited. Since the ILD will require international collaboration, there will be a fair number of groups involved in the enterprise. One of the first of these groups will be the International Lunar Survey Working Group (ILSWG), which will be tasked with sharing and disseminating scientific information among national participants. It is envisioned that the ILSWG will work within the framework of the International Space

International Lunar Decade 141

Exploration Coordination Group (ISECG), which will work with existing groups such as the National Space Science Data Center and the United Nations University. With the ISECG taking the lead in coordinating the activities of the groups, a long-range framework for developing the infrastructure on the Moon would be created. This infrastructure would include identifying sites for communication hubs, energy systems, and commercial and scientific operations. The groups would also develop a common standard for docking and landing, salvage, rescue, and working agreements on the myriad support services required to make a permanent return to the Moon possible. Another important aspect of the international collaborative effort will be to define working agreements, trial protocols, recognition of property rights, and the rights to access by national and commercial entities.

You can get some sense of the scope of what the ILD hopes to achieve by looking through the many development projects and milestones, some of which are listed here:

1. Establish fuel depots in low Earth orbit (Figure 7.3) and at Lagrange locations to extend travel capabilities beyond LEO.
2. Create a navigation and communications infrastructure that enables precise navigation in lunar orbit and on the lunar surface.
3. Establish tracking observatories capable of detecting asteroids and space debris.
4. Construct a research and gateway facility at the Earth–Moon Lagrange 2 station.
5. Establish a permanent beachhead manned habitat on the lunar surface via commercial participation.
6. Develop cube sat-sized spacecraft to enhance educational, commercial, and scientific opportunities.
7. Pursue research and development of in-situ resource utilization.
8. Encourage international participation in the development of new missions, new commercial investments, and new marketplaces for goods and services.

7.3 A NASA concept of an orbital fuel depot. Credit: NASA

The ILD is intended as a framework only – a living document that is updated and refined as the development of the activities needed to ensure a permanent presence on the Moon evolves. And establishing a permanent commercial manned presence is the primary goal of the ILD, the argument being that only by developing the Moon industrially will it be possible to reduce the costs and increase the reliability of space exploration. So the thinking of the ILD is long-term and doesn't end with boot prints on the Moon. The focus is on generating economic activity on the Moon and in *cis*-lunar space so the space economy can be expanded deeper into the Solar System: to Mars perhaps? One of the models the ILD group uses as a reference for international collaboration is the ISS. By partnering with each other, the various agencies involved in the development of the ISS should have been able to cut costs, which they did, although the average cost per astronaut day on orbit is around US$7.5 million, so cutting costs is a relative term in this case. But the reason the ISS was so expensive was because it was subject to a government-driven approach to design and development. It also didn't help that it was the Shuttle that did much of the heavy lifting to LEO and, at half a billion dollars a launch, building the ISS was always going to be an expensive proposition, collaboration or no collaboration. But, in 2015, with SpaceX bringing down the cost of launch to orbit dramatically and the focus being on commercially directed missions as opposed to government-directed ventures, the goals of the ILD may just have a chance of being realized.

The idea is that, by fostering international collaboration, the ILD will catalyze the path towards industrial development of the Moon and this in turn will lead to the creation of space goods and services that will result in the emergence of an economic model for off-world trade. Then, gradually, through competition, cost reductions will be realized, lunar activities will become more and more routine, and more investors will be persuaded that this lunar enterprise business might not be such a bad thing to invest in after all. And it's not as if the ILD model hasn't been put to the test before. Back in 1957, at the launch of the Space Age, the International Geophysical Year (IGY) was a collaboration of more than 60 countries that left behind a legacy of networks that continue to play a role in advancing science more than 50 years later.

So what must be done for the ILD to be given the green light? Well, first the ILD proposal must be approved by the Committee on the Peaceful Uses of Outer Space (COPUOS), which was established by the UN General Assembly in 1959. If COPUOS does not sanction the ILD, then the initiative is dead in the water, which is why the US and the other major spacefaring nations will need to support the ILD proposal. And if the ILD proposal is given the green light? What then? Well, then there's every reason to expect a manned landing on the Moon in the near future, and not just by the US, but by other nations and commercial entities as well. It might not be an easy sell because the ILD is proposing an investment of US$150 billion over 10 years, which is similar to the amount of money invested in the ISS over the same period of time. But that figure was so high because the ISS was a government-driven project. What could be done with US$150 billion in the era of commercially driven space development and in a time when launch costs are coming down significantly? Well, it's not unreasonable to assume that the number of people working on the Moon at the end of the ILD could be higher than the number working on the ISS (six) today by an order of magnitude. But, before we start daydreaming, the ILD has to become a reality and, for that to happen, COPUOS member states have to propose that the ILD is endorsed by the UN General Assembly. If that happens and the ILD is endorsed by the 59th COPUOS Session scheduled for 2016, then the return-to-the-Moon venture is in business and it will be a real tailwind.

SPACE RACE

Only three countries have ever sent astronauts into space: the US, Russia, and China. That's an elite group, and China intends to join another select club when it lands its taikonauts on the Moon. Over the years, the Chinese have methodically and incrementally taken bigger and more significant steps in space. Among those steps was China's first spacewalk performed by Zhai Zhigang in 2008 and the orbiting of the country's first space station – Tiangong – in September 2011. Most recently, all the talk was about the Jade Rabbit, also known as Yutu, a rover that landed on the Moon in December 2013. And, very soon, the Chinese may be in a position to set up camp on the lunar surface and the odds are good that they will do this before the US. If that happens, then the US may have to abdicate its role as leader in the manned-spaceflight business. Given the prospect of a red flag on the Moon, there have been calls for collaboration, but any teaming-up of the US and China would be technologically beneficial only to the Chinese and most US lawmakers have voiced their opposition to such a union on the grounds of China's human rights record and the fact that China's space program is tied directly to its military, which is at odds with American interests. Although the Chinese have expressed interest in being a part of the ISS enterprise, such participation was effectively deep-sixed in 2012 by US Congressman Frank Wolf who wrote a letter forbidding NASA entering into any such partnership. Here's an extract of Wolf's letter:

> "NASA should make clear that the U.S. will not accept Chinese participation in any station-related activities. The Chinese 'civilian' space program is directly run by the Peoples Liberation Army (PLA). I believe that any effort to involve the Chinese in the space program would be misguided, and not in our national interest."

So China is an outsider in the space community, which may strike some as a strange state of affairs when you consider that the US and the Soviet Union were able to bury the hatchet at the end of the Cold War and begin an era of on-orbit cooperation that started with NASA astronauts visiting Mir and continues today with American astronauts traveling to the ISS via the Russian Soyuz. So why can't the Chinese and the US get along? One reason is technological espionage: there have been a number of incidents over the years in which Chinese nationals have been accused of, and in some cases caught, smuggling sensitive information from NASA. Such actions do little for Sino-American relations.

So, while cooperation is off the table, is there a chance that China's lunar ambitions could lead to a space race that stimulates the US into recalibrating their exploration policy (such as it is) and accelerates the country's return to the Moon? Stranger things have happened. After all, the international perception is that manned spaceflight is something that the US does best, but images of a red flag and taikonauts wandering around Tranquillity Base might cause many people to reconsider that assessment. Some question whether the Chinese have the technological wherewithal to actually execute such a manned mission anytime soon because they are too far behind technologically. But the reality is that the Chinese have matched the achievements of the Russians and the Americans and the Chinese space program receives generous funding from a unified Chinese leadership. The US may have forgotten about Apollo, but the dream of taikonauts on the lunar surface is one that burns very brightly in Beijing. That's because landing on the Moon would represent not only an astounding national achievement, it would also go a long way to erasing two centuries of humiliation at the hands of the world's superpowers. And Chinese boot prints on the Moon would also be viewed as a powerful statement in power projection and space superiority. Emboldened by such an achievement and its position as the leading space power, what would there be to stop the Chinese from asserting territorial sovereignty on the site of their Moon base and being even more assertive?

"China right now is experiencing its Apollo years. China gets the funding its needs."
Joan Johnson-Freese, a professor at the US Naval War College

It's exactly that kind of scenario that could conceivably spark a space race and that could be the best thing that ever happened to fast-track back to the Moon. As China continues its stepwise approach to realizing a manned lunar landing, NASA's lack of strategic direction will eventually result in a loss of American leadership in space – a possibility that will become fact once taikonauts land on the Moon. But that event could be the galvanizing moment that NASA needs to refocus and redouble its efforts to returning its own astronauts to the lunar surface. Then again, some argue that a Chinese lunar landing won't matter because, by the time China lands on the Moon, NASA will be sending its astronauts to asteroids and preparing for manned missions to Mars. That could happen, in which case perhaps the space race will be between China and American companies such as Shackleton Energy Company?

ASTEROID

If you happened to be in the Chelyabinsk area in Russia on 15 February 2013, chances are you'll still have vivid memories of what happened that day. At 09:20 local time, a 17-meter-wide chunk of space rock entered Earth's atmosphere, traveling at around 70,000 kilometers per hour. Those in the Urals turned their attention skyward to witness Earth's largest meteor event in more than 100 years as the light from the bolide turned brighter than the Sun. Thanks to the object's high velocity and shallow angle of entry, the rock exploded in an air burst in the area near Chelyabinsk at an altitude of 30 kilometers. The explosion shattered windows, and the shock wave injured more than 1,000 people and damaged more than 7,000 buildings. It could have been much worse because, if the meteor had arrived at a different angle, it could have impacted the ground. All of a sudden, meteors were front-page news and many people were appalled at just how unprepared Earth is for events like these. While there are a number of programs that monitor large near-Earth objects (NEOs), there are plenty of large rocks up there that can bypass detection – the Chelyabinsk meteor being one of them. And, when it comes to stopping potential planet killers (see sidebar), we have nothing. All sorts of techniques have been suggested: everything from nuclear weapons to high-energy lasers, but none of these is operational. But, with the Chelyabinsk event, there may be a recalibration of priorities with a little more emphasis on space which means those advocating a return to the Moon may have a louder voice. So, while Chelyabinsk was a headwind for those affected directly by the event, the meteor may actually be a tailwind. For example, an argument could be made that one of the best places to locate a planetary defense observatory is the Moon. Such a venture, if approved, could fast-track a return to the Moon.

Tunguska

If you are a fan of the *X-Files*, you'll be familiar with the Tunguska event. At around seven in the morning on 30 June 1908, a shock wave obliterated 80 million trees and more than 2,000 square kilometers of forest. Wolves, bears, reindeer, and thousands of other animals were killed instantly. The explosion generated so much light that people 10,000 kilometers away could read their newspapers at night. The first scientists on the ground arrived in 1921 and pronounced the event to have been caused by a meteorite despite there being no crater. A more likely explanation was that a comet had been the culprit: one that probably weighed more than a million tonnes and was moving faster than 100,000 kilometers per hour. Had the comet arrived about four hours later, it would have wiped St Petersburg off the map. Will there be another Tunguska? Definitely. Perhaps another Tunguska will be needed to finally get countries to take the issue of planetary defense seriously.

8

Making the Moon Pay: The Economical and Logistical Viability of Boot Prints on the Moon

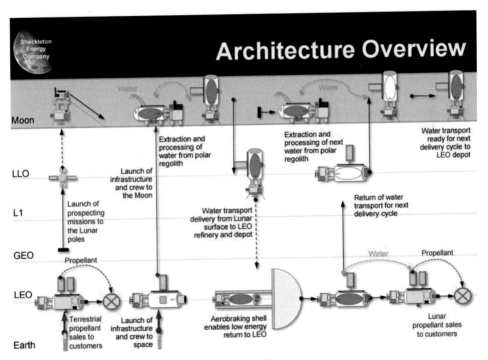

Credit: NASA

"In 1998, NASA sponsored a New Space Industries Workshop to peer into the future of space development. The final report of the workshop, titled *New Space Industries for the Next Millennium*, evaluated future space businesses such as tourism, space manufacturing, satellite services, space solar power, and more. The report makes fascinating reading. Today, though most of these businesses still do not exist, even in nascent form, in large part because the start-up costs for building the necessary space infrastructure are simply prohibitive."

Quote from "The Commercial Launch Industry Needs a Boost" by Douglas Jobes, The Space Review, *2 May 2005*

The era of "flags and footprints" missions is over. The next manned mission to the Moon will almost certainly be a commercially driven venture and that means the company or corporation financing the mission will want to make money. Shackleton Energy Company (SEC), Deep Space Industries, and the Chinese – they all want to go to the Moon and make money, but how do you make a convincing business case to mine the Moon? Well, the first step to take before you start calling investors is to make absolutely sure that you have the legal groundwork in place that allows you to do what are planning on doing. This means you probably need to convince the potential investor that you have ownership rights to whatever it is you want to mine. As discussed in Chapter 6, there is no entity that authorizes mining operations on the Moon and there are still some international legal issues that need to be ironed out before investors feel comfortable handing over billions of dollars. The next step is to employ experienced people and, on that count, the prospective mining companies score well. But these companies still face an uphill struggle in closing a business case because closing a business case for an actual business is very different from closing a business case for a concept. How do they pull it off? Well, you really need to be a bit of a wordsmith and be very careful in your choice of words when describing your prospective off-world enterprise. Phrases such as "growing momentum" and "exponential growth" are always guaranteed to catch the investor's eye, whereas "sustainable" is a word to be avoided because "sustainable" means stagnant and no investor is going to sink billions of dollars into a venture that is going nowhere. Another strategy is to convince the investor to make a leap of faith but not in one bold giant leap. Better to lay out a series of small leaps and bridges to those leaps. Once you've done that, you get to the real deal-breaker: access. Terrestrial business plans assume you can access your site of operations but, when you're planning on setting up shop on the lunar surface, this assumption can't be made because access is contingent on access to the Moon becoming affordable. But let's assume you can convince your investor that you will be able to gain access. What else does this business plan have to include? Well, there should be a solution to a problem. For example, SEC plans to process propellant on the lunar surface and make it available in low Earth orbit (LEO) for business being conducted there (Figure 8.1). That solves a problem because sending fuel from Earth is diabolically expensive. Sending propellant from the Moon also happens to be a value proposition because it is providing a service more cheaply than is currently available. Next is identifying the competition. The positive aspect of this is that the lunar companies have each found their niche, so the only competition is from the government and they tend to be on board with the commercialization of space anyway.

Now the investor will ask what's in it for him. What are the risk factors, how do you mitigate them, what are the markets, what are the alternatives, and what is the time horizon? This is where it gets a little awkward because a lunar business plan is based on

8.1 Bigelow's space station may be one of Shackleton Energy Company (SEC)'s clients in low Earth orbit (LEO). Credit: Boeing

assumption piled on top of assumptions piled on top of … well, you get the idea. So, at this point, the investors will nod their collective heads and start defining an exit strategy just in case this business plan goes down the toilet. Obviously, putting together a business plan, whether it be for a business here on Earth or on the surface of the Moon, is a lot more involved than the process described here but the intention is to provide a brief overview of the challenges faced by these budding lunar entrepreneurs. And, to provide further insight into how this lunar commercialization may evolve, it is worth taking a look at some of the industries that might be the most viable. We'll start with helium-3 (He-3). While there are many rare mineral and metals in the regolith, the resource that is the most valuable is He-3. He-3, along with regular helium, hydrogen, and nitrogen, has been deposited on the lunar surface by the solar wind for about four billion years. Analysis of the regolith indicates that the concentration of He-3 is between 20 and 30 parts per billion (ppb) in undisturbed soils but this figure may be as high as 50–60 ppb in the higher latitudes as a result of cold trapping. At the lower end of the scale, 20 ppb may not sound like much, but He-3 is worth about 50 times as much as gold (US$28 per gram of gold versus US$1,400 per gram of He-3). So just 100 kilograms of He-3 would be worth US$140 million. And those 100 kilograms of He-3 would provide sufficient fuel to run a 1,000-megawatt power plant for a year. Of course, getting all that He-3 won't be easy because it will require processing about two square kilometers of regolith down to a depth of more than two meters. And to do all that processing will require miner-processors which will have to be tended by maintenance and operations personnel. Based on studies performed by budding lunar-mining corporations such as SEC, it has been estimated that developing all the mining, refining, and processing capabilities and maintenance facilities would require an investment of around US$2.5 billion over a period of five years. Assuming a selling price of US$140 million per 100 kilograms for He-3, it might take a while for a company such as SEC to break even. And, since it is necessary to convince investors that they will get their money back – and then some – one of the key pieces of information will be to provide an accurate assessment of how much He-3 there actually is up there.

Studies that have extrapolated sample data returned by the Apollo missions indicate the 84,000 square kilometers of the *Mare Tranquillitatis* contains about 5,000 tonnes of He-3. That's an awful lot of He-3 and, with those sorts of numbers, you would think it would be an easy task to persuade an investor to mine the stuff but, before we look at the business case, let's take a closer look at how we might use this wonder gas. First, there is no question that Earth is in need of energy. We're running out of oil and all the talk is of the search for alternative energy sources, but why He-3 exactly? The main reason is that He-3 has high energy density in a fusion reaction, which means you only need a small amount of He-3 to provide a lot of energy. He-3 also happens to be very safe compared with, say, nuclear energy. But what about hydrogen and solar energy? Surely these are safe as well, aren't they? Yes they are. In fact, hydrogen (Figure 8.2) has long been marketed as an energy of the future because the gas is so widely available. It is such a clean energy source that Iceland has committed itself to running its entire energy infrastructure on hydrogen by 2020. The European Union (EU) has also climbed onto the hydrogen bandwagon by announcing its intention to become the first hydrogen-based superpower. On some levels, hydrogen makes sense. After all, it stores almost three times more energy per unit mass than petrol but the downside is that it needs about four times the volume to provide the same amount of energy. Another downside is that hydrogen is an energy carrier and not an energy source. What is meant by this is that hydrogen is generated by means of an

8.2 Liquid-hydrogen tank at Launch Complex 39A. Credit: NASA

endothermic process which translates into high operating costs. Not only that, but that endothermic process must be fueled by fossil fuels, so the gas isn't as green or clean as you might think. That's not to say we should give up on hydrogen as an energy source, it just means that hydrogen is not a long-term energy solution.

Another popular energy option is nuclear. This energy source relies on fission to generate energy and the fission process is one in which uranium is bombarded with neutrons to split the uranium atoms, thereby releasing energy. But the release of that energy is not very efficient due to the energy losses incurred by the use of turbines and mechanical machinery. And, while nuclear energy is considered fairly green, there is a cost associated with dealing with all that radioactive waste. A greener option is solar energy, which can be used to produce electricity via photocells, but there are still a number of thermal conversion challenges that must be solved before this energy source can be utilized on a commercial scale. So let's return to He-3. It's super-efficient and its energy potential is 10 times higher than recoverable fossil fuels and about twice as high as the uranium that is used in fast breeder reactors. Here's an example illustrating just how efficient this fuel is. One tonne of He-3 reacted with 0.67 tonnes of deuterium would generate about 10,000 megawatts of energy. To generate an equivalent amount of energy using oil would require 130 million barrels of the black stuff. At US$40 a barrel, this amount of oil would cost US$5.2 billion, so just one tonne of He-3 is worth more than five billion dollars. Sounds great, but we haven't discussed one very important subject: fusion (Figure 8.3). Can it be done? Probably (see sidebar).

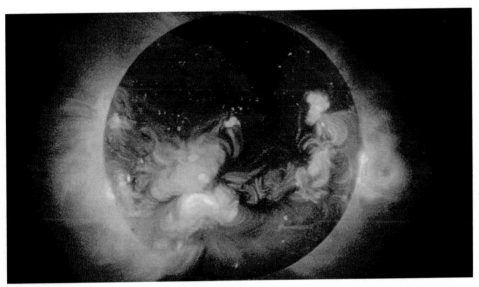

8.3 Solar flare. Credit: NASA

> *Helion Energy*
>
> Fusion power is the Holy Grail of nuclear physics. Imagine this: a test tube the size of your index finger can hold enough deuterium-fortified water to produce 18 megawatt-hours of electricity. Given this energy potential, it isn't surprising that there are research institutions around the world[1] trying to crack the fusion code, but why is it so difficult? The main reason is that it takes more energy to create potential fusion than is produced by the reaction. Another reason is that there are myriad variables – temperature, magnetic fields, colliding atom cores, plasma conditions – that must be in perfect balance for the process to work. The process works fine in the Sun's core, but achieving those conditions on Earth is a difficult nut to crack. But one company thinks it might be closer than most to figuring it out. Helion Energy has been in the business of building incrementally larger reactors in their attempt to fire two hydrogen atom cores at one another to create conditions similar to those found in the Sun. Key to their success is their Venti reactor, which is still being designed.

There are two options when it comes to fusion reactors. One type uses inertial confinement fusion (ICF) and the other, which is the most promising type for commercial applications, is magnetic confinement fusion (MCF). There is a lot of work going on to prove that a MCF reactor can work but, as of 2015, no system has broken even in the energy-production stakes. So you can talk all you want about how 17 square kilometers of the lunar surface would provide sufficient electricity to power a city of 10 million for a year, but it doesn't help the business case when your means of producing said energy doesn't exist! But our potential investor is patient, so he's interested in hearing about how the company is planning on extracting this miracle gas.

MINING

There have been a number of lunar-mining scenarios proposed over the years but the basic elements remain the same. Essentially, an automated mobile lunar rover would excavate regolith and electrostatically separate particles that are 50 micrometers in size, since these have the highest He-3 concentration. The soil that is left after this process – and there

---

[1] Some of the more prominent fusion research entities include the Massachusetts Institute of Technology, the National Spherical Torus Experiment in Princeton, General Fusion in Vancouver, and the Laser Megajoule in France. The world's biggest fusion project, ITER (originally an acronym of the International Thermonuclear Experimental Reactor and Latin for "the way"), located at Cadarache, France, which receives funding from 35 countries, reckon their US$50 billion facility can be successful by 2027.

would be a lot of it – would be dumped back onto the surface. The particles would be super-heated to more than 600°C to separate trapped gases which would be collected in tanks. The rover tasked to undertake the excavation might look something like the Mark II Miner (Figure 8.4) (see Table 8.1).

Once collected and compressed, the gases would be processed by a condensing station which would cool the gases to liquid form while the helium would be super-cooled to isolate the He-3 isotope. The energy for all these processing activities would most likely come from solar power and fusion. The bulk of the He-3 would be shipped to Earth as portrayed in the movie *Moon*. Feasible? Definitely, but very tough. For starters, the lunar soil is very, very different from terrestrial soil. Lunar soil has no moisture, the surface friction is much higher on the Moon than on Earth, and the rock strength is much higher, which is part of the reason the Apollo 15 astronauts broke their drill stem (Figure 8.5).

Assuming a beefed-up Mark II Miner can deal with the very tough regolith, the vehicle would mine the soil using a rectilinear strategy by roving back and forth across the lunar surface or it might adopt a spiral mining approach, which is similar to open pit mining on

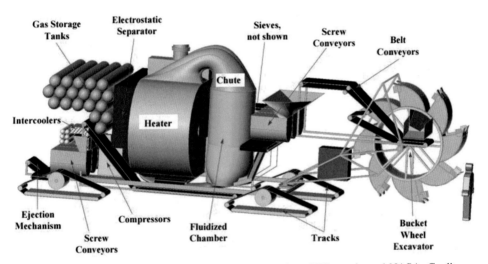

8.4 The Mark III Miner as developed by the University of Wisconsin and NASA. Credit: University of Wisconsin/NASA

Table 8.1 Mark II Miner design specifications.

Annual He-3 collection rate (10 ppb)	33 kg
Mining hours per year	3,942
Excavation rate	1,258 tonnes per year
Excavation depth	3 m
Mark II Miner speed	23 m/hr
Area excavated	1 km²/year
Processing rate	556 tonnes/year

8.5 Dave Scott picking up the drill which broke during Apollo 15. Credit: NASA

Earth. Then, once the He-3 is processed, there needs to be a way of shipping it to Earth. Here there are two options. One is to liquefy the He-3 and the other option is to leave it as a gas. Liquefying obviously reduces the volume of the gas but the costs of liquefying and maintaining the super-cold temperatures during transit may be prohibitive, so He-3 will likely be transported as a gas. One scenario for transporting the product to Earth will be to first transfer the payload to lunar orbit and from there to LEO before making the last leg of the journey via a conventional commercial shuttle.

Sounds fairly straightforward, doesn't it? But, from the investor's perspective, there are an awful lot of assumptions. It's possible a commercial company could pull this off but perhaps a better business case could be made by an international consortium. Such an organization could use current energy treaties as leverage, arguing that He-3 represents a clean energy that will help reduce greenhouse gases in accordance with the Kyoto Protocol of 1997. Still, without the technology being available to actually utilize He-3, it's still a tough sell to an investor. Even if the investor can feel comfortable with all the assumptions, there is still the question of financial viability which basically equates to maximum return on investment. To calculate this return once again requires some assumptions to be made because the investor needs to know how expensive this will be. The first cost is lugging all that mining equipment from Earth to the surface of the Moon. Once that's done, the second objective is maintaining the Moon base, which will likely be staffed by four or five maintenance personnel. The third cost item will be the mobile miners required to actually extract the He-3 and then the company must shell out for the separation and condensation systems to actually separate the He-3 ready for transport. Finally, there must be a transportation architecture in place to ferry the product back to Earth, and all this is assuming that the fusion nut will be cracked in the near future. Numbers have been crunched in an attempt to answer these questions, but the outcome is vague at best. For example, the University of Wisconsin estimated that 51 kilograms of mining material would have to be transported for every one kilogram of He-3 produced. That sounds like a lot, but the value of He-3 should more than cover that capital investment. Should. Then there is the energy payback to consider. Studies performed by NASA's Solar System Exploration Committee reckon mining He-3 would result in a payback ratio of more than 200, which sounds like a promising number to a potential investor, but these are estimates. A more rigorous estimation is difficult given the assumptions that must be made so, if you happen to be a budding He-3-mining enterprise, how do you convince your investor? Perhaps you could play the monopoly card: whoever gets there first will have a stranglehold on He-3 production. It's one option.

So how might this He-3-mining business play out? Well, assuming fusion is demonstrated by the end of the 2020s, there is a good chance that the first step towards realizing the mining of lunar He-3 will be in place: a manned base. Once fusion has been demonstrated, the challenge of persuading investors becomes a lot easier and they will likely pony up the billions of dollars required to establish the mining infrastructure on the lunar surface. By this time, the Chinese may have their own Moon base and may be establishing their own He-3-mining enterprise, which will mean that China may be competing with the US in the fusion-energy stakes. It's difficult to predict how the development of a He-3-mining operation will play out but, once fusion becomes a reality and once a lunar-based mining operation is up and running, He-3 will almost certainly be recognized as a revolutionary agent for change not only in the terrestrial energy sector, but also in the business of manned spaceflight. Which country or commercial entity will take advantage of the mother lode of He-3? Again, it's difficult to predict because the price tag of getting involved in such a venture is huge and, with economic returns sketchy until fusion has been proven, it would be a brave investor who stepped up to the plate. Perhaps it would be easier to pitch a mining business on the back of resource for which a market does exist. Platinum, perhaps?

PLATINUM

Metallic asteroids have been leaving their marks on the Moon for eons. And among the metals contained in those asteroids is platinum and platinum group metals (PGMs). One of the proponents of mining PGMs is Dennis Wingo, whose hypothesis was neatly summarized by Jeff Foust in a *Space Review* article:

"The Moon ... would seem to be an unlikely place to find PGMs: the collisional process that formed from the Moon left it mostly devoid of heavy metals. However, Wingo makes an ingenious case for finding PGMs on or near the lunar surface, in the form of debris from asteroid impacts. While conventional wisdom has argued that impacts of large asteroids would vaporize most of the impactor, modern computer modeling has shown that a significant fraction of an asteroid impacting the Earth would survive in some form. In fact, some major sources of PGMs on Earth, such as Sudbury in Canada and sites in South Africa, have been linked to asteroid impacts. The Moon's lower gravity would mean slower impacts, making it more likely that significant portions of asteroids could survive. PGMs mined from those impacts could meet the fuel-cell needs of the Earth for centuries; the mining process would, in turn, also generate other metals like iron and nickel that could be used for settlements on the Moon and beyond."

"Moonrush" by Jeff Foust, The Space Review, *16 August 2004*

The 2015 price of platinum is around US$35,000 per kilogram and, given the metal's industrial usefulness, there is definitely a market, especially in the US, which must import more than 90% of its yearly demand. But that figure is based on purified platinum and the platinum on the Moon is anything but purified. So any company interested in the lunar platinum business would have to set up a mining infrastructure similar to those hoping to mine He-3. The bonus for those extracting platinum is that a lunar-based platinum-enrichment facility could also be tweaked to enrich other PGMs, so the business wouldn't be reliant on just one metal. Again, any such enterprise would have to crunch the numbers to see whether such a business was viable. If they can, such a business may have found a niche market.

PROPELLANT

So He-3 is a long shot in the near term because fusion has yet to be proven and platinum may work but we still don't know the concentrations of PGMs so the business case for mining these metals is a little flimsy. But what about propellant? This is where SEC think they have a strong business model. SEC's goal is to establish the company as the leading off-world energy company by providing rocket propellant to those operating in LEO and on the Moon. Thanks to spacecraft such as the Lunar Reconnaissance Orbiter (Figure 8.6),

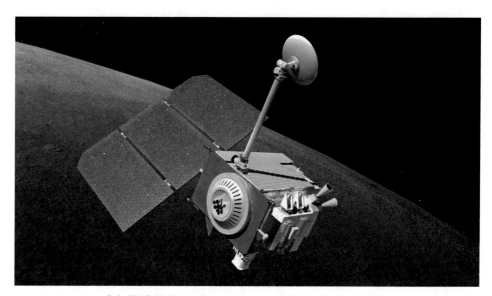

8.6 NASA's Lunar Reconnaissance Orbiter. Credit: NASA

we know that the Moon contains water, and lots of it. NASA reckons there could be billions of tonnes of the stuff and, since water can be transformed into liquid oxygen (life-support consumables) and liquid-hydrogen propellants, SEC reckon they have a very strong business case by meeting the increasing propellant demand of the rapidly expanding off-world economy.

And SEC has a better chance of succeeding than most because their funding will not be sourced from government coffers or even traditional investment. Instead, SEC plan on funding their initiative via a combination of mining and energy-related sources. One example of how SEC are going about their business is collaborating with industries that have very close parallels with mining on the Moon, one such business being the oil and gas industry. In March 2015, SEC signed a Memorandum of Understanding with Norwegian company Zaptec to assess how oil and gas technology can be adapted to be used on the Moon. The agreement made perfect sense because Zaptec has a patent on technology that can transform high voltages in small devices thereby dramatically reducing material usage: this technology can be applied to plasma drilling, which is one of the cornerstones of the mining operations that SEC envisage on the lunar surface.

But let's go back to SEC's business case. We know there are a number of entrepreneurial space companies hard at work trying to develop a LEO-based economy. Bigelow Aerospace is planning on sending his BA-330 expandable modules into orbit as soon as he has access to a manned crew vehicle and SpaceX may be one of the companies that provide that transportation. Then there are those companies that have already been mentioned who are aiming to set up shop on the Moon. In parallel with these efforts, launch companies are striving to bring down launch costs to reduce development and operating costs of those companies working off-world. This is important to SEC because, to kick-start their business, they first have to lug all their infrastructure to the Moon. Once they are established on the Moon, SEC can take advantage of the lower gravity gradient and develop a flexible, module-based, and reusable transportation system that can further develop their core business of propellant supply. And SEC have a very real chance of realizing their business having assembled an extraordinary team of experts in the fields of engineering, mining, and aerospace, economists, and space policy specialists. In addition to their very capable team, SEC have myriad relationships with various levels within the international space industry and academia. But mining lunar ice to make rocket fuel? Can it be done? Well, SEC are convinced it can be done because the technology to do this has been around for a while, although it still has to be tested on the Moon. The exact details of how the water and ice will be mined will be finalized once SEC have completed a series of manned and unmanned surveys of the Moon. Once this phase is complete, it is likely that automated rovers will strip mine the ice and water-laden regolith, and the water will be extracted by heating the crushed soil using solar power. Water will then be loaded onto rovers and flown to LEO where it will be transferred to an orbital fuel-processing facility.

At the fuel-processing facility, the water will be first turned into gaseous hydrogen and oxygen, and then into liquid hydrogen and oxygen. This rocket fuel will then be stored in cryogenic containers to be used to fuel spacecraft. One advantage of the process is that it is very clean and the processing is relatively simple. And, for those concerned about the environmental impact of all that strip mining, SEC are planning on returning debris to the craters. Another advantage is the business case, which – at least on paper – is very profitable. At its very core, SEC's business provides a simple solution to a very common problem and that is the insane cost of getting fuel to LEO. But, once SEC have installed all their infrastructure on the Moon, they will be able to reduce the exorbitant cost of space travel in orbit and thereby provide a healthy return on investment which will keep the investors happy. Even assuming SpaceX can realize their cost advantage and bring down the cost of launching payloads to LEO to US$1,000 a kilogram, that's still an awful lot to pay for, say, a liter of water. That's because anything launching from Earth has to use a lot of fuel to get to LEO whereas anything launched from the surface of the Moon needs comparatively very little fuel. All of sudden, that same liter of water may cost as little as US$100 or US$200. Still more than you pay for your Perrier but, in the spaceflight arena, a couple of hundred dollars for a liter of water is a steal. And, once you have water and fuel in situ, it becomes a lot easier to live off the land. Still, there is some way to go before this becomes a reality. The inventory of volatiles on the lunar surface has to be determined, the purity of those volatiles has to be assessed, and their accessibility has to be measured. It will be helpful to know how the water ice is distributed in the regolith and also how a company goes about locating the ore and excavating it. More importantly, how much will all this cost? So there is a lot of technology demonstration that is needed before we have scenes similar to those portrayed by Hollywood in the movie *Moon* (see sidebar), but the good news is that there is plenty of momentum targeted at providing answers to these questions.

Moon: Hollywood's take on lunar commercialization

Moon plays out in a time when mining the lunar surface is an accepted way of earning a living. Sam Bell, played brilliantly by Sam Rockwell, is the sole engineer of a He-3-mining facility owned by Lunar Industries. Keeping him company is a slightly condescending computer called Gerty. As he's nearing the end of his three-year tour of duty, Sam crashes one of the company's harvesters and wakes up in the infirmary to find he has a doppelganger. Is he hallucinating or is he losing the plot after three years isolated from human contact? Gerty informs Sam that a rescue team has been deployed from Earth to fix the harvester and Sam is to remain in the base. Sam is suspicious, leaves the base anyway, and opens the hatch of the crashed harvester to find another Sam. Things don't add up. He asks Gerty what is going on and Gerty replies that the other Sam is Sam Bell. The Sams are clones. Every three years, a new one is awakened and imprinted with the same memories. The two clones put their heads together and come up with a plan. The first Sam climbs back into the harvester while the newer Sam launches back to Earth just as the rescue team arrive. The credits roll as a voiceover of a radio show announces that Sam Bell will be testifying against Lunar Industries over the cloning controversy.

8.7 The MX-1. Credit: Moon Express

In addition to SEC, California-based Moon Express are also interested in extracting water from the lunar regolith and their first step towards cashing in utilizing lunar in-situ resources will be sending their MX-1 lunar lander to the Moon in late 2015. Weighing in at 600 kilograms, the MX-1 (Figure 8.7) will deliver 60 kilograms to the lunar surface. Developed as part of the Google Lunar X-prize (a competition to land a vehicle on the Moon, travel 500 meters, and send photos back to Earth), the MX-1 is the first in a series of robotic vehicles (Moon Express has already partnered with NASA to build a lunar lander), with its primary goal being to dig around and collect lunar soil to see how much water and rare elements are on the surface. If all goes well, the MX-1 could be the catalyst for Moon Express to deploy more missions to further explore and evaluate the lunar surface with an eye for further development.

Appendix I

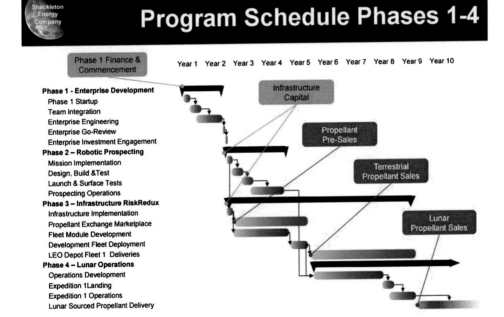

Shackleton's plans

Appendix II: International Lunar Decade Goals

1. Develop a plan for the systematic international study of lunar resources including mineral, geographic, and orbital resources, assigning responsibilities to countries, firms, and organizations based on their capabilities determined through a competitive process.
2. Develop an internationally shared and accessible database of lunar resources, including procedures for updating and accessing the database and preserving the security of information.
3. Develop procedures for communications between different research groups, outposts and business incubators, tourist bases, and other facilities operating on the Moon.
4. Establish a competitive process for providing access to specific firms seeking to exploit specific resources in specific locations on the Moon such that they can exploit the resources according to the principles set forth in Article 11, paragraph 7 of the Moon Treaty.
5. Develop a Lunar Development Fund that coordinates the international development of key enabling technologies required to advance exploitation of the resources of the Moon as well as their use in pre-commercial, pilot applications, and business incubator facilities that may be located in specific regions of the Moon.
6. Launch competitive calls open to research institutes and research oriented firms from around the world to develop technologies and operational consortia to develop required technologies for the exploitation of lunar resources.
7. Develop specific programs, incorporating a competitive process that encourages excellence, to increase the space research and space technology development capabilities of states that have had limited opportunity thus far.
8. Develop long-term financing mechanisms that can lead to the development of commercial-scale resource extraction, energy production, transport, and other facilities on the Moon and in cislunar space.

9. Provide for mechanisms for conflict resolution including an appeals process to higher authorities.
10. Develop an organizational structure for a Lunar Development Corporation to support the fulfillment of the above objectives within the general principles set forth in the Moon Treaty particularly as relates to the international regime referred to in Articles 11.
11. Secure sufficient funding from participating governments and private sources such that the above objectives can be effectively fulfilled during the course of the International Lunar Decade.

Epilog

"First, you go to the moon before you go to Mars. You're not going to go to Mars before going back to the moon. You need to establish a goal to go to the moon and do that first and have a program laid out for an effective way to do it, but they're not doing that right now and I think that's really key to exploration."

George W.S. Abbey, former director of NASA's Johnson Space Center, in an interview with the International Business Times, *December 2014*

"If we return to the moon just for science and exploration then activities will be limited by the amount of money our nation is willing to devote. But, if we establish a sustainable, economically viable lunar base then our science and exploration will be limited only by our imagination."

Astronaut Ron Garan

The Moon is where the action is going to be in the 2020s and 2030s and beyond. It will serve as a vital stepping stone and test bed that will enable us to venture deeper into space and eventually all the way to Mars. Between July 1969 and December 1972, there were a few astronauts who spent some time on the lunar surface. That was more than 45 years ago and the time has long since passed for a return. But, this time, a manned mission will not be a "flags and footprints" venture – this time, the goal will be to learn to live in deep space and to test the myriad technologies required to enable an eventual manned mission to the Red Planet in the very distant future. Given the state of current national space policy in the US, the next boot prints on the Moon may be Chinese or they may belong to a commercial enterprise – it's difficult to say. What we do know is that NASA's proposal to send a crew to capture an asteroid has received less than a lukewarm reception and we also know that NASA's chief, Charlie Bolden, is adamant that there will be no manned mission to the Moon during his tenure as administrator. So a commercial return to the Moon seems likely, perhaps led by a consortium that follows a multi-participatory strategy akin to the collaboration that has been so successful demonstrated by the partners of the International Space Station (ISS).

Why New Space? Well, the rise and rise of the commercial sector is thanks in part to the industry following its own agenda and motivations, which is a path no government can follow. Since New Space companies have their own agenda, they are much more agile in setting and achieving goals. Companies like Shackleton Energy Company (SEC) don't have to spend inordinate amounts of time discussing policy objectives, nor do they have to spend countless and often fruitless hours engaged in seemingly endless and protracted dialog debating the next destination. SEC doesn't have to ask itself why they are sending people into space and they don't have to worry about how the US, or any other nation for that matter, can use manned spaceflight to advance its interests. SEC know why they are going to the Moon and that reason is to establish a propellant-production facility to supply those engaged in *cis*-lunar and low Earth orbit (LEO) activities. Simple.

Now to that manned mission to Mars. Most spacefaring nations agree that a manned mission to Mars should be the ultimate long-term objective and most agree that the only way to achieve such a monumental goal is to first return to the Moon. There have been whole books written that advocate "living off the land" as a way to realize a manned mission to Mars but very few of the technologies needed to achieve this have been field tested. And then there is the life-support question. Do we have a bioregenerative life-support system capable of sustaining a crew for three years? We don't. Which means we have to lug tonnes and tonnes of life-support consumables, which means more mass, which means more cost, which in turn means the whole enterprise becomes too expensive. And, even if we did have a bioregenerative life-support system that had been tested and tested again, what about the radiation issue? Yes, radiation again. The topic has been discussed at length in this book but it needs to be mentioned again because radiation rules out any potential manned mission to Mars:

> "Radiation shielding is the most overlooked feature of proposed interplanetary vehicles. NASA and space industry mission planners consistently underestimate the radiation hazards on a trip to Mars, particularly from GCRs [galactic cosmic rays] and thus minimize the shielding to protect against this exposure. The conventional wisdom states: 'NASA cannot afford to shield against radiation because the enormous mass penalty will make a Mars mission too expensive.' However, a truly safety conscious

approach insists 'NASA cannot afford NOT to shield effectively against radiation, despite the mass penalties.' It is time for NASA and the space industry to face up to radiation exposure as a major concern for crew health and for their ability to carry out a successful mission and to protect the crew against it."

Dr. M. Cohen

Thanks to the Radiation Assessment Detector (RAD) that was carried along with the Curiosity rover to Mars, we have a much better idea of the deep-space radiation environment. For example, we know thanks to an article published in *Science* in May 2013 that a 360-day return trip to the Red Planet will result in a radiation exposure of 662 ± 108 millisieverts and we know that 95% of that radiation will be from galactic cosmic rays (GCRs) which are almost impossible to shield against. So how do we provide our Mars-bound crew with protection equivalent to the protection provided by Earth's atmosphere? Well, one way to do it would be to build a spaceship with a hull one meter thick. Or perhaps we could use water. This subject is often brought up in discussions about how to protect Mars astronauts against killer radiation: just put them in a storm shelter surrounded by water. Job done. Not quite. Lets engage in some hypothetical mission design for a moment. Let's assume the manned Mars vehicle is the same size as the Mars500 cylinder – that is, the spacecraft measures about 3.5 meters in diameter by 20 meters in length. Now let's propose we use water to shield the vehicle. We know that to provide protection equivalent to 5,500 meters of altitude on Earth would require a hull five meters thick, but this is a budget mission, so the engineers have decided that a one-meter-thick hull will have to do. With one meter of water shielding around the hull, the hull would be 5.5 meters in diameter and 22 meters in length. Now let's calculate the volume of shielding. This done by calculating the difference between the two cylinders: $22\pi(5.5^2) - 20\pi(3.5^2) = 1,321$ cubic meters. Now we know a cubic meter weighs 1,000 kilograms, so we have to get 1,321,000 kilograms into space. How do we do that? Even if we have the Space Launch System (SLS) which may (if it is built) have a maximum payload capacity of 130,000 kilograms, you would need more than 10 launches and it would be pointless exercise anyway because the radiation shielding isn't enough to keep the astronauts safe. For that, you need a hull that is at least five meters thick. You do the math on that one.

So returning to the Moon will buy scientists some valuable time to sort out the radiation problem … and the life-support challenge … and the in-situ resource utilization … well, you get the idea. So, as the ISS winds down, commercial enterprise will head to the Moon. Commercial contracts will be signed to deliver propellant to LEO and to *cis*-lunar space. Resources will be mined and government procurement may leverage commercial investment which may result in cost reductions. And, while these activities are taking place, experience is being gained that can help those who eventually strike out for Mars. So the lunar surface will become not just a destination, but part of a commercially led drive to develop and establish a permanent presence on the Moon. As a technical and logistical training ground, the lunar surface will provide the ideal platform to advance the planetary technology needed for eventual Mars settlement. And, because this will almost certainly be a commercially driven venture, it can adopt an agile approach with a higher tolerance for risk than traditional government missions. No doubt about it, the surface of the Moon is the only logical place for the next giant leap.

Index

A
Acute radiation syndrome, 6
Aerocapture, 18–21
Alzheimer's, 17–18
Apollo, 4, 10, 11, 19, 20, 28, 30, 32–34, 42, 45, 48, 52, 61, 75, 77, 85, 86, 132–135, 138, 144, 150, 153, 154
Appendectomy, 110
Asteroid Redirect Mission (ARM), 50–52, 54, 131
Atlas V, 75, 80, 83
Automation, 44

B
Bigelow aerospace, 57, 58, 66, 71–73, 129, 130, 149, 157
Bioregenerative life-support system (BLSS), 87–89, 91–93
Bone loss, 2, 13–17
Business model, 156

C
Cataracts, 5, 8–13, 17, 18, 114, 121
China, 28, 35–40, 50, 52, 55, 127, 129, 143, 144, 155
Closed-loop life support, 86
Collaborative pathways, 50
Commercial justification, 71
Contour Crafting, 98–100
Cryogenic propellant, 157

D
Delta IV, 80, 82
Descent and landing, 18–25, 135

E
Entry, 2, 18–25, 33, 39, 53, 60, 85, 94, 111, 144
European Space Agency (ESA), 28, 35, 36, 40–46, 49, 55, 64, 67, 115, 139, 140

F
Falcon heavy, 80, 81, 84, 85
Fusion, 37, 59, 125, 150–156

G
Galactic cosmic radiation, 17
Gallbladder, 112, 113
Gallstones, 112
Genetic testing, 114
Golden spike, 74–77

H
Helion energy, 152
HTP *See* Hypersonic transition problem (HTP)
Human physiology, 62
Hypersonic transition problem (HTP), 23

I

ICP *See* Intracranial pressure (ICP)
ILD *See* International Lunar Decade (ILD)
Infrastructure development, 61, 66, 68, 141, 155
In-situ resource utilization, 66, 100–106, 141
International Lunar Decade (ILD), 140–143
Intracranial pressure (ICP), 12, 122

K

Keravala, Jim, 68, 69, 130
Kessler syndrome, 139, 140

L

LRO *See* Lunar reconnaissance orbiter (LRO)
Lunar assets, 58, 72, 129
Lunar prospecting, 7, 50, 59, 68, 101, 102, 111, 112, 114, 148
Lunar reconnaissance orbiter (LRO), 67, 156
Lunar station, 28, 35, 36, 66, 143, 149

M

Mining, 58–60, 67, 68, 70–72, 126, 128–131, 135, 136, 148, 149, 152–158

O

Omics, 114
OpenLuna, 58, 73–74

P

Platinum, 59, 155–156
Pollution, 135–136
Pre-emptive surgery, 111–113

Project horizon, 63
Project M, 96
Property rights, 58, 71, 72, 129–131, 141

R

Regolith, 8, 38, 41, 44, 49, 51, 55, 61, 63, 66, 68, 71, 74, 76, 92, 93, 99–105, 135, 149, 152, 153, 157–159
Rogozov, L., 109–111
Robotics, 93–98

S

Salvage, 126, 132–134, 141
SEC *See* Shackleton Energy Company (SEC)
Shackleton Energy Company (SEC), 58–60, 66–70, 85, 126, 129, 130, 144, 148–150, 156–159
Shielding, 3, 8, 11, 17, 80, 100
Shimizu corporation, 70–71
SinterHab, 43, 44
Solar particle event (SPE), 2, 5, 6, 8, 9
Space Launch System (SLS), 47–51, 80, 84
SPE *See* Solar particle event (SPE)
STP *See* Supersonic transition problem (STP)
Sustainable space enterprise, 77
Supersonic transition problem (STP), 23

V

VIIP *See* Visual Impairment Intracranial Pressure (VIIP)
Vision for Space Exploration (VSE), 8, 47, 49
Visual Impairment Intracranial Pressure (VIIP), 12, 13, 122
VSE *See* Vision for Space Exploration (VSE)

Made in the USA
Middletown, DE
14 December 2022

18296067R00108